MANUAL FOR THE HOME AND FARM PRODUCTION OF

ALCOHOL FUEL

Stephen W. Mathewson

COVER PHOTOGRAPH: Experimental 3 column still built to check operating parameters of a new plate design.

—Courtesy ALCOHOL TECHNOLOGY CORPORATION

Printed in U.S.A.

Second Edition
January 1980

CONTENTS

FIGURES AND ILLUSTRATIONS

Figures and Illustrations continued:

INTRODUCTION

Production of alcohol fuel can be approached on many levels, from costly installations designed to supply the fuel needs of large farms, to small, economical systems that enable even city-dwellers to save considerable amounts of gasoline. However, making alcohol fuel takes knowledge, equipment and work. There is no "free lunch" and solving the energy problem will not be easy. Nevertheless, it _is_ possible if we make the effort.

There are many things to consider before deciding on a "personal" alcohol fuel facility. Some of them are economic, such as how much fuel is needed, how much the feedstock will cost, how much the equipment will cost, and how much labor will be involved. Other considerations are more philosophical. For example, how much is motor fuel really worth? The trouble and expense of setting up a manufacturing and refining plant to produce fuel is not for everyone. Some will simply choose to drive less or pay the ever increasing fuel costs. However, there are others who, for one reason or another, cannot afford to be dependent on the whims of oil producers or the capriciousness of government fuel allocation. To a farmer, for example, with crops to be harvested and an empty tank in his tractor, an alcohol production capability not only makes sense, it could mean survival.

In the following pages, I have tried to present a complex subject in as clear and lucid a manner as possible. Ethanol production involves chemistry, biology, engineering, and a certain amount of common sense. Although this book contains the basic information necessary to produce and use alcohol fuel, it should be considered only a primer.

Stephen W. Mathewson

Chapter 1 AN OVERVIEW

ALCOHOL FUEL

There is nothing new in the use of alcohol as a motor fuel. In 1872, when Nikolaus Otto invented the internal combustion engine, gasoline was not available. Ethyl alcohol at 180-190 proof was the specified fuel. The model "T" Ford was designed to run on the available crude gasolines, alcohol, or any combination of the two.

Alcohols in general and ethanol, in particular, make excellent motor fuels. The reason alcohol fuel has not been fully exploited is that, up until now, gasoline has been cheap, available, and easy to produce. However, crude oil is getting scarce, and the historic price differential between alcohol and gasoline is getting narrower.

Currently there is a big push to find and develop alternative sources of energy so that dwindling reserves of crude oil and other fossil fuels may be conserved. As Edward Teller, one of this country's leading physicists points out: "No single prescription exists for a solution to the energy problem. Energy conservation is not enough. Petroleum is not enough. Coal is not enough. Nuclear energy is not enough. Solar and geothermal energy are not enough. New ideas and developments will not be enough by themselves. Only the proper combination of all of these will suffice."

Alcohol fuel can be an important part of the solution, but it is by no means a panacea. If all of the available agricultural surplus were converted to ethanol, alcohol would supply less than 5% of our motor fuel needs. Add the possibility of converting cellulose residues to ethanol and general biomass to methanol, and the most optimistic total falls short of 10% of our present needs! However, this is a very important 5 or 10% because it can be renewed each year, and each gallon of alcohol produced will save a gallon of oil.

USES OF ALCOHOL FUEL

One very important fact about alcohol fuel should be stressed. Alcohol is an excellent alternative motor fuel for gasoline engines. It is not a suitable alternative for home heating or for essentially stationary power requirements. The production of alcohol consumes energy. Exactly how much depends on the feedstock (raw material) and the efficiency of the distillation process. In a small operation, it would not be uncommon to expend 30-40,000 Btu per gallon of ethanol. It would be more sensible, in a home heating situation, to use whatever fuel you would use to run the still directly rather than using it to produce alcohol. The real advantage of alcohol is that it can be burned in the millions of existing vehicles with little or no modification. Alcohol fuel should only be considered for the jobs it can do best.

OTHER ALTERNATIVE FUELS

This book is about the small scale production of ethanol for use as a motor fuel. However, before becoming committed to ethanol, there are other alternatives that should be considered.

The first that comes to mind is methanol, or "wood" alcohol. Like ethanol, methanol is a viable substitute for gasoline, and it can be produced from a wide variety of renewable biological resources. Methanol, however, is not as easy to produce on a small scale.

The simplest and oldest method of producing methanol is by the destructive distillation (pyrolysis) of wood. The process is nothing more than heating the wood residues in a "dry" distillation apparatus and collecting the methanol at the other end. As such, the process requires relatively simple equipment and should be suitable for small scale production. The problem, then, is the fact that along with the methanol a considerable amount of impurities are produced that include acetone, acetic acid, and a number of other substances. These by-products are difficult and expensive to remove, and, if left in the methanol, they will quickly corrode an engine. Simply put, the small scale production of methanol by destructive distillation requires a large enough plant to justify the equipment and energy necessary to remove the impurities. If you happen to have a large source of suitable hardwood and are prepared to make the necessary capital investment, methanol production by this method might be considered.

Other processes for producing methanol from renewable resources, such as hydrogen and carbon monoxide, or conversion of cellulose and biomass, also exist. Again, the problem is that these methods are only feasible on a very large scale.

Methane gas has also been considered as a motor fuel. Methane is generated, for example, by the action of bacteria on manure. The problem here is that any methane production facility must be large enough to justify the equipment and energy required to compress the gas for storage. Also, methane has a very low heat value (energy content per unit of weight) and engine conversion is necessary. Methane is better suited to stationary power requirements than for use as a motor fuel. Still, if you have a situation where a large amount of manure or other suitable biomass is available, methane generation should be considered.

Natural gas, propane, and butane are also possible motor fuels. However, since all of them are basically petroleum related, they cannot be considered as renewable resources.

Much research has been done to find better processes for separating water into hydrogen and oxygen in order to obtain the hydrogen for use as a fuel. To date no process has been developed that does not consume more energy than can be returned when the hydrogen is burned.

Aside from alcohol and, perhaps, methane, there seems to be no other suitable alternative fuel that can be made from renewable resources and utilized in existing motor vehicles. Other means of powering vehicles, such as electricity, involve the development and production of completely new vehicles. What seems to be needed is a vehicle that can utilize a wide variety of fuels such as coal, wood, alcohol, gasoline, kerosene, corn cobbs or whatever might be available--for instance, something similar, to the 1897 Stanley Steamer!

But, in the meantime, ethanol is the best solution for a motor fuel from renewable resources that can be produced easily on a small scale.

Chapter 2
BASIC FUEL THEORY

CHEMICAL COMPOSITION

Alcohol and gasoline, despite the fact that they are from different chemical classes, are remarkably similar. Gasoline is mostly a mixture of "hydrocarbons". Hydrocarbons are a group of chemical substances composed exclusively of carbon and hydrogen atoms. This is a very large chemical class containing many thousands of substances. Most of the fuels we use such as coal, gasoline, kerosene, fuel oil, butane, propane, etc. are chiefly hydrocarbons. Referring to Figure 2-1, the simplest member of this group is methane which consists of a single carbon atom and

Figure 2-1: CHEMICAL STRUCTURES

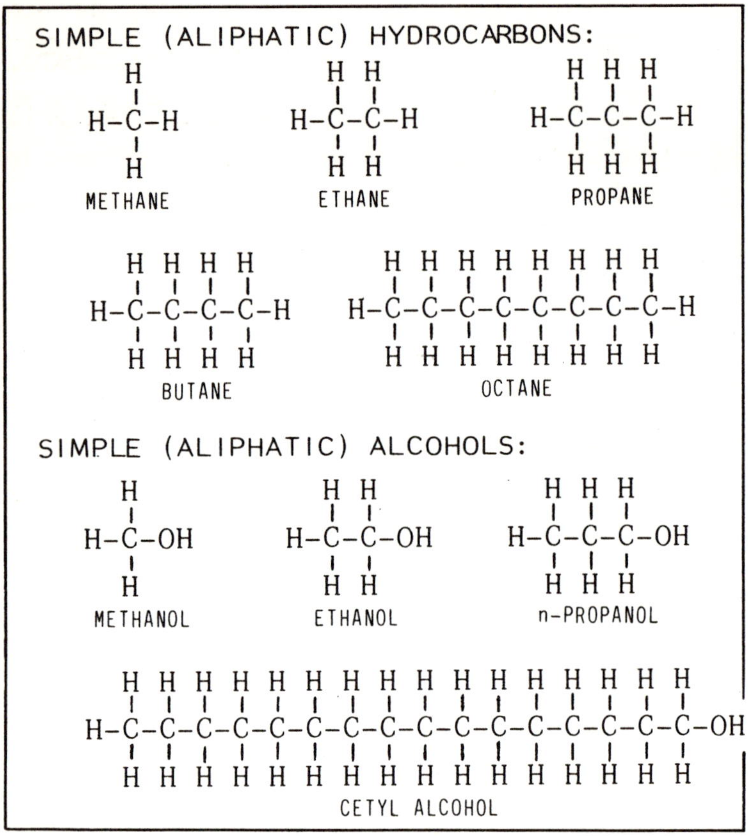

four hydrogen atoms. Next comes ethane with two carbons and six hydrogens. Propane has three carbons and butane has four. The substances just named are gases under ordinary conditions. As we add more carbons to the hydrocarbon molecule, the chemicals formed become liquids: pentane, hexane, heptane, octane and so on. As we continue with even more complex molecules, the substances get progressively oilier, waxier and finally solid.

Alcohols can be thought of as hydrocarbons in which one of the hydrogen atoms has been replaced by a "hydroxyl group" which consists of a hydrogen atom bonded to an oxygen atom. Thus methane becomes the simplest alcohol, methanol. Ethane becomes ethanol, propane becomes propanol and so on. Like hydrocarbons, there are many alcohols of ever increasing complexity.

COMBUSTION PROPERTIES

One of the most important properties of a fuel is the amount of energy obtained from it when it is burned. Referring to Figure 2-2, note that the hydrocarbon octane, which represents an "ideal" gasoline, contains no oxygen. In comparison, all of the alcohols contain an oxygen atom bonded to a hydrogen atom in the hydroxyl radical. When the alcohol is burned, the hydroxyl combines with a hydrogen atom to form a molecule of water. Thus, the oxygen

Figure 2-2: PHYSICAL PROPERTIES of ALCOHOL and GASOLINE

	TYPICAL REGULAR GASOLINE	OCTANE*	METHYL ALCOHOL	ETHYL ALCOHOL
Chemical Formula	Complex	C_8H_{18}	CH_3OH	C_2H_5OH
Molecular Weight	Complex	114	32	46
Heating Value (Btu/lb)				
High Value	20,250	20,570	9,770	12,780
Low Value	19,000	19,080	8,640	11,550
Latent Heat of Vaporization (Btu/lb)	140	141	474	361
Specific Gravity (@60°F)	0.745	0.702	0.796	0.794
Stoichiometric Ratio	15:1	15.1:1	6.45:1	9:1
Boiling Temperature (^{o}F)	100-400	258.2	148.5	173.3
Octane Number (Research)	80	100	106	106
Energy of Stoichiometric Mixture (Btu/ft^3)	94.8	95.4	94.5	94.7

*Can be considered as "ideal" high-test gasoline

contained in the alcohol contributes nothing to the fuel value.

The relative atomic weights of the atoms involved are: hydrogen, 1; carbon, 12; and oxygen, 16. Since methyl alcohol has an atomic weight of 32, half the molecule cannot be "burned" and does not contribute any fuel value. As expected, methanol has less than half the heat value (expressed in Btu/lb) of gasoline. Ethanol, with 35% oxygen, is slightly better with 60% of the heat value of gasoline.

If the heating value of methyl and ethyl alcohol were considered alone, they would appear to be poor choices as motor fuels. However, other redeeming qualities such as "latent heat of vaporization" and anti-knock values make alcohol fuels superior, in some ways, to gasoline.

When a fuel is burned, a certain amount of air is required for complete combustion. When the quantity of air and the quantity of fuel are exactly balanced, the fuel/air mixture is said to be "stoichiometrically" correct. Again referring to Figure 2-2, the stoichiometric ratio for gasoline is 15:1 or 15 pounds of air for each pound of gasoline. The figures for methyl and ethyl alcohol are 6.45:1 and 9:1 respectively. On a practical level, this means that to burn alcohol effectively, the fuel jets in the carburetor must be changed or adjusted to provide 2.3 pounds of methanol or 1.66 pounds of ethanol for each 15 pounds of air.

Referring to the last entry in Figure 2-2, an interesting fact is that if we provide the correct stoichiometric mixture and then compare on the basis of the energy (in Btu's) contained in each cubic foot of the different fuel/air mixtures, the fuels are almost identical: gasoline 94.8 Btu per cubic foot; methanol 94.5 and ethanol 94.7! This means that gasoline and alcohol are about equal in what is called "volumetric efficiency" when burned in a correctly adjusted engine.

VOLATILITY

Another important quality in a motor fuel is "volatility", or the ability to be vaporized. As previously noted, methyl alcohol contains less than half the heat value of gasoline and ethyl alcohol contains only about 60%. The next higher alcohol, propyl alcohol with three carbon atoms, contains only 26.6% oxygen and thus about 74% of the heat value of gasoline. It is apparent that the more complex the alcohol, the closer its heat value comes to that of gasoline. Cetyl alcohol (Figure 2-1), for example, contains only about 6.6% oxygen and thus has about 90% of the heat value of gasoline. However, this alcohol is a solid wax! It can't be conveniently vaporized and mixed with air in an engine and so is useless as a motor fuel. Consequently, in considering alcohol fuels, a compromise must be made between heat value and volatility.

Closely related to volatility is a quality called "latent heat of vaporization". When a liquid is at its boiling point, a certain amount of additional heat is needed to change the liquid to a gas. This additional heat is the latent heat of vaporization, expressed in Btu/lb in Figure 2-2. This effect is one of the principles behind refrigeration and the reason that water evaporating from your skin feels cool.

Referring to Figure 2-2, gasoline has a latent heat of about 140 Btu/lb; methanol, 474 Btu/lb; and ethanol, 361 Btu/lb. In an engine, vaporization of the gasoline fuel/air mixture results in a temperature drop of about 40° Farenheit. Under similar conditions, the temperature drop for ethyl alcohol will be more than twice that of gasoline, and for methanol the drop will be over three times as great. These temperature drops result in a considerably greater "mass density" of the fuel entering the engine for alcohol as compared to gasoline. The result is a greatly increased efficiency for alcohol fuels. To visualize why, remember that at a given pressure, the amount of space a gas occupies is directly proportional to the temperature. For example, if one pound of a gas fits into a certain container at a given pressure and the temperature is cut in half, the container will now hold two pounds of the gas at the same pressure. In an engine, a stoichiometric mixture of methanol and air would be over three times colder than the same gasoline/air mixture. This means that there is now over three times (by weight) as much methanol in the cylinder. Now, even though methanol has only half the heat value of gasoline, the net gain in "volumetric mass efficiency" is over three times. So, for example, if the gasoline/air mixture in a given engine cylinder produces 100 Btu on each stroke, the same engine would produce 150 Btu per stroke with methanol. This power gain due to increased volumetric mass efficiency is the primary reason for the popularity of methyl alcohol as a racing fuel. With ethanol the effect isn't quite as dramatic, but the greater heat value partially offsets the lower latent heat. Overall, this power increase with alcohol fuels considerably mitigates the liability of low heat value.

However, the increased cooling due to latent heat sometimes creates a problem in an engine converted to run on alcohol. Once vaporized, a certain amount of heat is required to keep the fuel from condensing back to the liquid state before it reaches the cylinder. To accomplish this, an engine is designed to provide this heat to the intake manifold. Alcohol, bcause of its greater latent heat, requires more heat than gasoline. This is one of the reasons that racing engines have short path manifolds and multiple carburetors. The shorter the distance the fuel must travel to the cylinder, the less chance of condensation and fuel distribution problems. On a practical level, most engines that have been converted to alcohol supply enough heat once they are warmed up. The main problem, as with high performance racing engines, is in starting a cold engine. This problem and the related fuel distribution problem will be discussed later in more detail.

OCTANE RATINGS

If a certain fuel is burned in an engine in which the compression ratio can be varied and this ratio is gradually increased, a point will be reached when the fuel will detonate prematurely. This is because as a gas is compressed, heat is generated. If the explosive fuel/air mixture in an engine cylinder is compressed enough, the resulting heat will cause it to detonate. Since gasoline engines are designed so that the mixture is detonated by the spark plug at the beginning of the downward movement of the piston following the compression stroke, preignition or "knock" occurring during the compression stroke is undesirable. Indeed, severe knock can quickly overstress and destroy an engine.

Since greater compression ratios in an engine

mean increased power per stroke and greater efficiency, the ability of a fuel to resist premature detonation is a desirable quality. The "octane" numbers assigned to fuels are based on the pure hydrocarbon, octane, which is considered to be 100. At the other end of the scale, n-heptane is considered to have an octane rating of zero. The octane number of an unknown fuel is based on the percentage volume of a mixture of octane and n-heptane that matches it in preignition characteristics. In practice, these tests are conducted in a special test engine with variable compression. As noted in Figure 2-2, alcohols have a relatively high anti-knock or octane rating. As noted in Figure 2-3, alcohols have the ability to raise considerably the octane ratings of gasolines with which they are mixed. The effect is greatest on the poorer grades of gasoline. A 25% blend of ethanol and 40 octane gasoline will have a net increase of almost 30 points! This increase is one of the major advantages of "gasohol". The ability to increase octane rating means that: (1) a lower (therefore cheaper) grade of gasoline can be used to obtain a fuel with a certain octane

Figure 2-3: OCTANE INCREASE of ALCOHOL/GASOLINE BLENDS

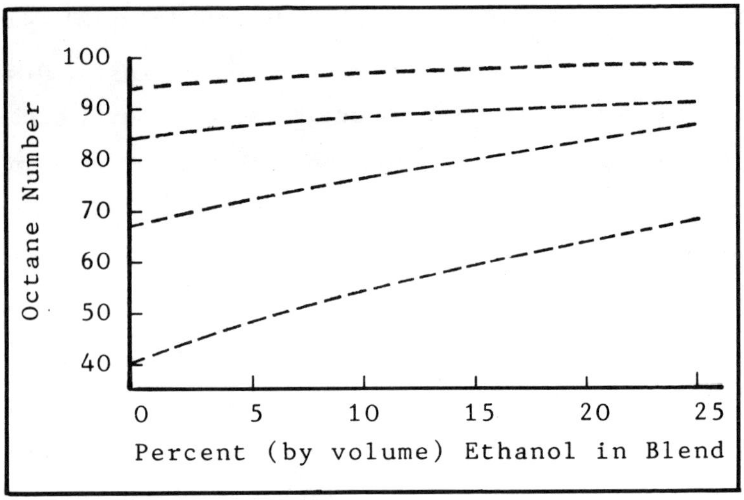

rating; and (2) the use of traditional pollution producing anti-knock additives such as tetraethyl lead can be eliminated. The addition of about 10-15% ethanol to unleaded gasoline raises the octane rating enough so that it can be burned in high compression engines that previously could not use unleaded fuel. This use of ethanol is not new, of course, because ethanol was the original gasoline additive for increasing the octane rating. The term "ethyl" used to describe a high-test gasoline comes from ethyl alcohol, not tetraethyl lead!

WATER INJECTION

During World War II, the military made extensive use of water injection in high performance piston aircraft engines. Later, water injection was used by both civilian and military jet aircraft to provide extra thrust, principally on takeoff. Even today, water injection systems are available that can be installed in automobiles. The fact is, that within certain limits, these systems actually do increase power. Referring back to Figure 2-2, note that the latent heat of vaporization for gasoline is about 140 Btu/lb and for ethanol about 361 Btu/lb. Water has a latent heat of about 700 Btu/lb! Therefore, if a little water is injected into the carburetor in the form of an ultra-fine mist, the latent heat of the water will cool the charge and increase volumetric efficiency. In addition, when the charge is fired in the cylinder, the water will turn to high-pressure steam and provide additional power due to the pressure exerted by the steam. There are definite limits, however, to the amount of water that can be injected. Too much will cause excessive cooling and misfiring.

The use of water injection with a gasoline fueled engine requires a separate metering and injection system because water and gasoline do not mix. Ethanol and water, however, do mix and the benefits of water injection can be had simply by adding the desired amount of water to the alcohol in the fuel tank.

EXHAUST COMPOSITION

In theory, a hydrocarbon fuel when burned should produce only water and carbon dioxide (CO_2) as exhaust gases. Carbon dioxide, of course, is completely non-poisonous being the gas we exhale when we breathe, the bubbles in carbonated beverages, and the gas plants turn back into oxygen during the photosynthesis cycle.

However, such ideal combustion rarely occurs even in the most perfectly adjusted engine. What is actually produced is a large amount of poisonous carbon monoxide (CO) and other complex (and undesirable) emissions arising from impurities like sulfur and additives such as lead or phosphorus.

Pure alcohol when burned under ideal conditions also produces, in theory, only carbon dioxide and water. Again, in practice, varying amounts of carbon monoxide are also produced. However, the amounts of carbon monoxide are usually much lower than with gasoline. In addition, alcohol fuel will contain no sulfur and no additives, and will not produce the related, undesirable combustion by-products. Pure alcohol fuels are extremely clean burning.

Many studies have been made to determine whether alcohol/gasoline blends have any positive effect on emissions. In general, the data show that no great changes occur in blends of 20% or less. What happens is simply that in a 10% alcohol/gasoline blend, for example, about 10% of the gasoline emissions are replaced with alcohol emissions. Since alcohol does burn considerably cleaner, the amount of emission improvement is proportional to the amount of alcohol in the blend.

Pure alcohol, as an anti-pollution fuel, would easily meet and exceed all emission requirements without the need for exotic and costly exhaust plumbing and catalytic converters. With alcohol blends, the chief advantage would be in the use of ethanol to replace lead and other undesirable compounds used to raise the octane number.

ENGINE PERFORMANCE-STRAIGHT ALCOHOL

Having looked at a few of the basic factors which influence the performance of fuels in an engine, let us now examine some actual engine tests. Figure 2-4 is a plot of 198 proof (99%) ethyl alcohol as compared to gasoline. "Mean Effective Pressure" in the graph is a direct indication of the power produced. The increased mean effective pressure (M.E.P.) of alcohol at all mixture ratios is the most noticeable difference between the two fuels. This increase in M.E.P. is due mainly to the greater volumetric efficiency that results from the high latent heat of vaporization of ethanol and the resulting greater mass density of the fuel/air mixture.

Figure 2-4: ENGINE PERFORMANCE of ETHANOL vs GASOLINE

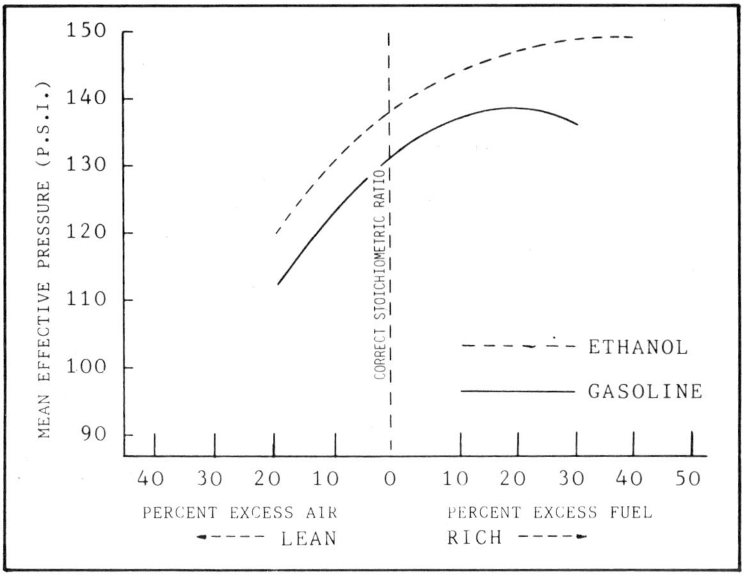

Note that the M.E.P. of ethanol increases with mixtures having up to 40% excess fuel, whereas for gasoline, the maximum pressure is reached at 20% excess fuel. It would seem that to achieve maximum power from an alcohol-burning engine there would be a temptation to burn very rich mixtures. Fuel economy aside, it should be noted that the rich mixtures necessary to obtain maximum M.E.P. are accompanied by incomplete burning of the fuel and the resultant lowering of overall thermal

Figure 2-5: HORSEPOWER COMPARISON of ETHANOL vs GASOLINE

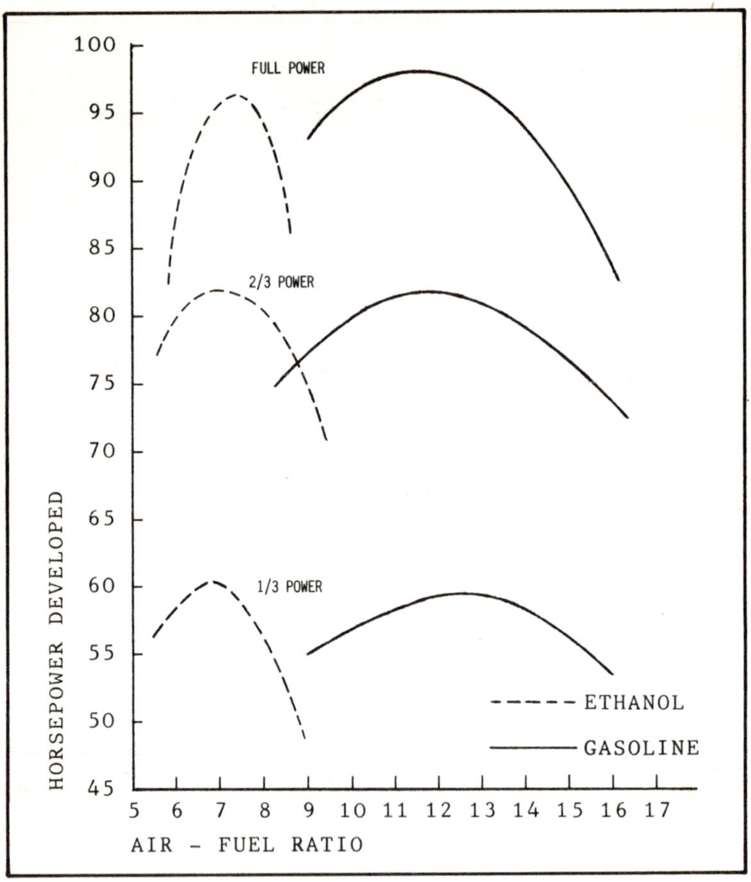

efficiency. The lean limits for alcohol and gasoline, therefore, are about the same, and both fuels develop maximum thermal efficiency at about 15% excess air. With mixtures leaner than 15% both fuels loose thermal efficiency.

Figure 2-5 compares engine horsepower and air/fuel ratios for ethanol and gasoline in a six cylinder engine. The fuels in this case were 190 proof (95%) ethanol and "regular" gasoline having a specific gravity of 0.745. In the tests, air was supplied to the intake manifold at a constant 100° Farenheit, and the carburetor needle valve was adjusted to provide the desired fuel/air ratios. The 2/3 and 1/3 loads were established by adjusting the throttle to give the same manifold pressure for both fuels.

The smaller air/fuel ratios for ethanol in comparison with gasoline are evident. In this test with the air supplied at the same temperature for both fuels, the correct fuel/air mixture should produce about 2% more power from gasoline than ethanol. However, alcohol, with its greater latent heat, requires more manifold heat to remain completely vaporized. In another test where this additional heat was supplied, the correct alcohol/air mixture gave 8.6% more power with ethanol! Note also that the test depicted in Figure 2-5 was run with alcohol that contained 5% water. This benefit of water injection probably inflated the alcohol power results to a certain degree. However, the main point illustrated is that the two fuels are remarkably similar in performance in a correctly adjusted engine.

ENGINE PERFORMANCE-ALCOHOL BLENDS

Although alcohol blends can be made from both ethanol and methanol, the primary interest seems to be in the direction of ethanol. Methanol and gasoline have a limited miscibility (mixability) while ethanol and gasoline can be mixed in all proportions. Economic reasons also dictate the interest in ethanol since it is more readily made from renewable resources. In addition, ethanol is a slightly superior motor fuel alternative under most conditions.

Economics aside, a major advantage of blends is that up to a certain concentration (somewhere between 10 and 20%) they can be used with absolutely no modification of the engine.

Many studies on how the various blends affect engine performance are contradictory. The recent "Two Million Mile" test in Nebraska, claims slightly higher fuel economy. Other tests claim a slight decrease. Some tests claim slightly better emissions, others claim no significant change. In relation to power output, the tests are equally ambiguous. However, when all the data is sifted, the overall conclusion is that in the areas of fuel economy, emissions, and performance there just isn't any real difference.

Figure 2-3, as discussed under Octane Ratings, illustrates another major advantage of alcohol blends, namely the ability of alcohol to raise the anti-knock quality of the gasoline with which it is mixed. This means, of course, that lower, cheaper grades of gasoline can be used to obtain a fuel with the desired octane rating, and the use of pollution producing additives can be eliminated. This is a significant advantage from the economic standpoint because the manufacture of high-octane blending stocks is expensive. Also, as previously mentioned, it is possible to raise the octane rating of unleaded gasoline so that it can be used in engines that previously required high-test leaded gasoline.

Alcohol blends do have one relatively minor drawback. The presence of even small amounts of water in the blend will cause a portion of the alcohol and gasoline to separate. At room temperature, less than 1% water can do the damage. As the temperature is lowered, amounts as small as .01% can cause separation. However, various substances such as benzene (benzol), acetone, and butyl alcohol can be added to the blend to increase water tolerance. Closed fuel systems, now in use, prevent moisture from forming inside the gas tank. Oil companies, given the proper incentive, could dry out their storage facilities and pipelines. Also, extensive use of alcohol blends over the past 50 years is ample evidence that the problem can be solved.

Chapter 3
UTILIZATION OF ALCOHOL FUELS

METHODS OF UTILIZATION

Alcohol fuels may be utilized in three basic ways: as a blend with gasoline; as a straight, unblended fuel; or as an alcohol/water mixture in an injection system. Each method has certain advantages and disadvantages.

ALCOHOL BLENDS

Alcohol blends have the advantage that up to a 10, 20 or even 25% concentration of alcohol may be used without modification to the engine. The actual concentration that may be used varies with each engine type, but generally a four-cylinder engine will tolerate a stronger blend than a six or eight. Small single-cylinder engines, such as lawn mowers, can often be run on pure alcohol by merely adjusting the mixture control screw. Even with larger engines, slight modification such as adjusting the carburetor and, perhaps, advancing the timing a little may allow the use of blends in the 25-40% range. If you are producing your own blend, you have the advantage of being able to use the cheapest gasoline available and ending up with a good, high octane fuel.

The disadvantage is that the alcohol you use must be perfectly dry. As will be discussed in the section on distillation, the highest concentration of alcohol that can be achieved by ordinary methods is 190 proof or 95%. In order to blend the alcohol with gasoline, the remaining 5% water must be removed. Several methods of removing this residual water will be discussed in Chapter 12, and one of them may be practical for you. However, drying the alcohol does require a separate manufacturing step and the expenditure of additional energy.

PURE ALCOHOL

The advantages of burning relatively pure 80-95% alcohol are several. First of all, because the drying step is unnecessary, you should be able to produce the fuel for less than the cost of gasoline. Secondly, there will be little, if any performance penalty, and by leaving 5-15% or more water in the alcohol you also gain the benefits of water injection. The only disadvantage is the trouble and expense of modifying your engine(s) to burn alcohol and the lack of dual-fuel capability.

The principal engine modification is the enlargement of the carburetor jet(s). If you are a reasonably competent mechanic, you should be able to do the job in a couple hours at a very small cost.

In addition to the carburetor jets, there is also the problem of cold starting. As mentioned earlier, alcohol has a higher latent heat of vaporization than gasoline and requires more manifold heat to keep the mixture in the vapor state. With most engines there will be no problem that can't be solved by installing a higher temperature thermostat since the engine runs fine as soon as it is warmed up. However, the engine will be difficult to start, especially in cold weather. The easiest solution to this problem is simply to start the engine on gasoline and, after it has warmed up, switch to alcohol. To accomplish this, merely install a small gasoline tank located, perhaps, under the hood and a selector valve mounted in some convenient location near the driver.

It is my experience that it is also desirable to replace the automatic choke with a manual control. Also, switching back to gasoline prior to shutting down the engine will aid in restarting. A more complex solution to this problem would be to install a priming pump and manifold heater glow plugs similar to those found on diesel engines. Other alternatives are to preheat the fuel or squirt an easily volatized liquid such as pentane into the carburetor. The addition of about 8% pentane directly to the alcohol in the fuel tank will also solve starting problems in below zero weather.

Another problem, also related to latent heat, is that of fuel distribution. Larger engines are more likely to encounter this problem than small ones. What happens is that there is insufficient heat to keep the fuel vaporized and some of it liquefies before it reaches the outer cylinders. This causes misfires and general poor performance. Simple solutions include insulating the intake manifold or installing a higher temperature thermostat. Heating the fuel before it enters the carburetor also helps, as does heating the intake air. The ultimate solution is, usually, to install multiple carburetors and a short-path manifold. However, you are likely to encounter this problem only in engines that are, by some design fault, prone to the same poor fuel distribution with gasoline.

It must be stressed that, although most engines are easily converted to alcohol, each engine is different. Some people have been able to successfully run Volkswagens and Hondas on alcohol merely by adjusting the jets and playing with the timing a little. Others, especially those with older, ultra-high performance V-8 engines, have had to resort to more extensive modifications.

Alcohol engine modification is a relatively "re-discovered" field. To the best of my knowledge, there is no comprehensive information on the modification of specific modern engines, and there are no manufacturers making conversion kits. With all the interest in alcohol fuels, however, this should change in the near future. At present, though, it seems that modifications must be made on an individual basis.

For those with special engines, there is good news and bad. Turbocharged engines present no special conversion problems once the jets, etc. have been enlarged. Alcohol and turbochargers then work very well together. Fuel injected engines are another matter. To convert them can be a simple adjustment of the metering pump, or it can be virtually impossible. Very little data seems to be available in this area.

DIESEL ENGINES

Contrary to the opinion of most "experts", diesel engines can be run on pure alcohol. The main problem is in the lubrication of the injectors. This is solved by the addition of 5-20% vegetable oil (or other suitable lubricant) to the alcohol. It is also possible to make a diesel "gasohol" with up to 80% alcohol. Since alcohol and oil will not mix when water is present, both the alcohol and the oil must be anhydrous. Different engines may also require adjustment of the metering pump for optimum performance. Diesel engines, especially turbocharged diesels, may also be run with an alcohol/water injection system as described later.

ENGINE MODIFICATION

The following are some specific guidelines to assist in the modification of a carburetor. Remember that there are many different types and makes of carburetors, and that a certain amount of experimentation will be necessary.

First, of course, you will have to remove the carburetor from the engine, clean it, and disassemble it to a point where you can remove metering jet(s). This will involve removing the air horn from the float valve and disconnecting any linkage. Next you must locate the main jet (or jets on a multi-throat model). Most carburetors have removable jets. They are almost always brass and are threaded into place.

With the jet removed, the next step is to measure its diameter. This is best done with a micrometer. You will want to enlarge the area of the jet about 27% for ethanol and 40% for methanol. Suppose, for example, your jet is .054" in diameter. The formula for the area of a circle is 3.14 (pi) times the square of the radius. The radius is half the diameter, so we multiply .027 x .027 x 3.14 to get an area of .002289 square inches. Multiply this times 1.27 (for a 27% .enlargement) and we get .00291 square inches. Working the formula in reverse we get a diameter of .06087 inches. This is close to a #53 drill which is slightly too small. Since it is easier to enlarge a hole than to make one smaller, a wise choice for the first trial in this instance would be a #53 drill.

Carefully drill out the jet, reassemble the carburetor, and reinstall it on the vehicle. The vehicle should then be run on alcohol as a test. Start the engine and slowly enrich the mixture (using the idle screw adjustment) until the engine starts to stall. Then adjust the idle until the roughness evens out. Take the vehicle for a short test drive, and then pull the spark plugs. If the tips are white, the mixture is too lean, and the main jet will have to be further enlarged. If they are wet, the mixture is too rich, and you have made the jets too large. In addition, if the mixture is too lean, the engine will backfire and miss. It will also burn the valves if left in this condition. On the other hand, if the hole is too large, the mixture will be too rich and you will waste fuel. It may be necessary to make several trials before the perfect jet size is found for your particular engine. In addition to the main jet, some carburetors will also require a slight enlargement of the idle circuit jet. This is accomplished in the same manner as above except that a smaller percentage of enlargement will usually suffice. Note that this modification isn't always necessary. Often merely backing out the idle adjustment screw will be enough.

If the engine still doesn't run properly, there are several other things you can try such as advancing the timing a little, disconnecting the vaccum advance line, and closing the spark plug gaps a little. If you want to go the whole route, you can increase the engine's compression by milling the head and installing high compression pistons because the alcohol's high anti-knock qualities will allow compression ratios to 10:1. Finally, if you have an engine where it is impossible to modify the carburetor, for one reason or another, (an excess of emission "plumbing", for example) you can usually replace your carburetor with an earlier model. Usually, the older the carburetor, the easier it is to convert. Also, it is possible to purchase adjustable jets for many carburetors, or your carburetor may already have such jets. Adjustable jets make it easier to change from alcohol to gasoline and vice versa should the need arise.

ALCOHOL INJECTION

Alcohol injection is a third alternative for the utilization of alcohol fuel. It is similar to water injection except that alcohol or an alcohol water mixture is injected into the engine. Since the water/alcohol injection mixture ratio can be as low as 50/50%, first run product from a simple still can be used. This is a considerable saving because most of the energy used in alcohol production is expended in the distillation stage to obtain 95% alcohol. Another advantage is that engines with an injection system still retain complete dual fuel capability. Finally, alcohol injection can be used with fuel-injected, turbocharged, and even diesel engines.

Figure 3-1 is a schematic of a simple injection system. The alcohol/water mixture is contained in a separate tank and is fed, under a couple pounds pressure, to a misting nozzle located at the throat of the carburetor. The mixture is

Figure 3-1: BASIC INJECTION SYSTEM

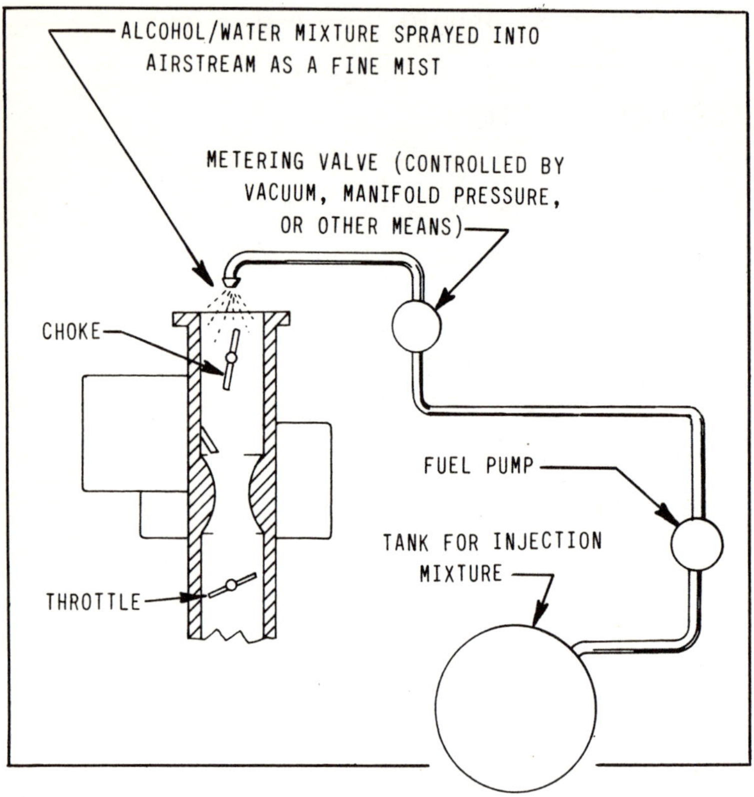

metered into the carburetor airstream where it mixes with the air and is taken into the engine. There are many ways of metering the alcohol/water mixture. For example, it can be done by mechanically linking a metering valve to the throttle. Other methods include using combinations of vacuum and/or manifold pressure. Whatever system is used, the

Figure 3-2: INJECTION of TURBO ENGINES

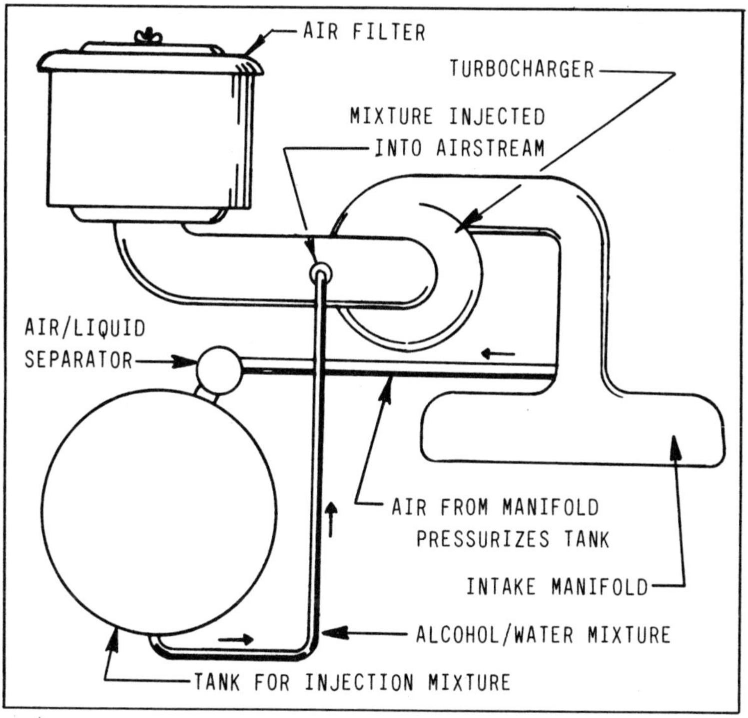

metering system should work in parallel with the throttle. That is, the flow of alcohol/water mixture should increase as the load increases. Figure 3-2 diagrams a similar system for turbocharged engines. This is an extremely simple system. The alcohol/water tank is pressurized by bleed air from the compressor on the turbocharger. The mixture is metered into the turbocharger airstream by an orifice.

The size of the orifice is determined by individual engine requirement. The metering system operates with the turbocharger. As the boost increases, more pressure is supplied to the tank and, thus, more mixture to the engine.

On a diesel tractor rated at 125 horsepower and consuming $8\frac{1}{2}$ gallons of fuel per hour, the injection system produced the same power with only six gallons of diesel fuel and two gallons of a 50% alcohol/water mixture. This is an overall saving of 6% in fuel consumption and a saving of almost 30% in diesel fuel. Other benefits include trouble-free, automatic operation, increase in available power, lowering of engine operating temperatures, and prolonged engine life.

Chapter 4
ETHANOL PRODUCTION -GENERAL DISCUSSION

RAW MATERIALS

Ethyl alcohol may be made by the fermentation process from three basic types of raw materials, called "feedstock".

The three basic types of feedstock are:

(1) SACCHARINE (sugar containing) materials in which the carbohydrate (the actual substance from which the alcohol is made) is present in the form of simple, directly fermentable six and twelve carbon sugar molecules such as glucose, fructose, and maltose. Such materials include sugar cane, molasses, sugar beets, fruit (fresh or dried),

citrus molasses, cane sorghum, whey and skim milk.

(2) STARCHY MATERIALS that contain more complex carbohydrates such as starch and inulin that can be broken down into the simpler six and twelve carbon sugars by hydrolysis with acid or by the action of enzymes in a process called malting. Such materials include corn, grain sorghum, barley, wheat, potatoes, sweet potatoes, jerusalem artichokes, cacti, manioc, arrow-root, and so on.

(3) CELLULOSE MATERIALS such as wood, wood waste, paper, straw, corn stalks, corn cobs, cotton, etc., which contain material that can be hydrolyzed with acid, enzymes or otherwise converted into fermentable sugars called glucose.

MANUFACTURING STEPS

The basic steps in the manufacture of fuel ethanol from each of the three groups is outlined in Figure 4-1. It can be seen that

Figure 4-1: MANUFACTURING STEPS

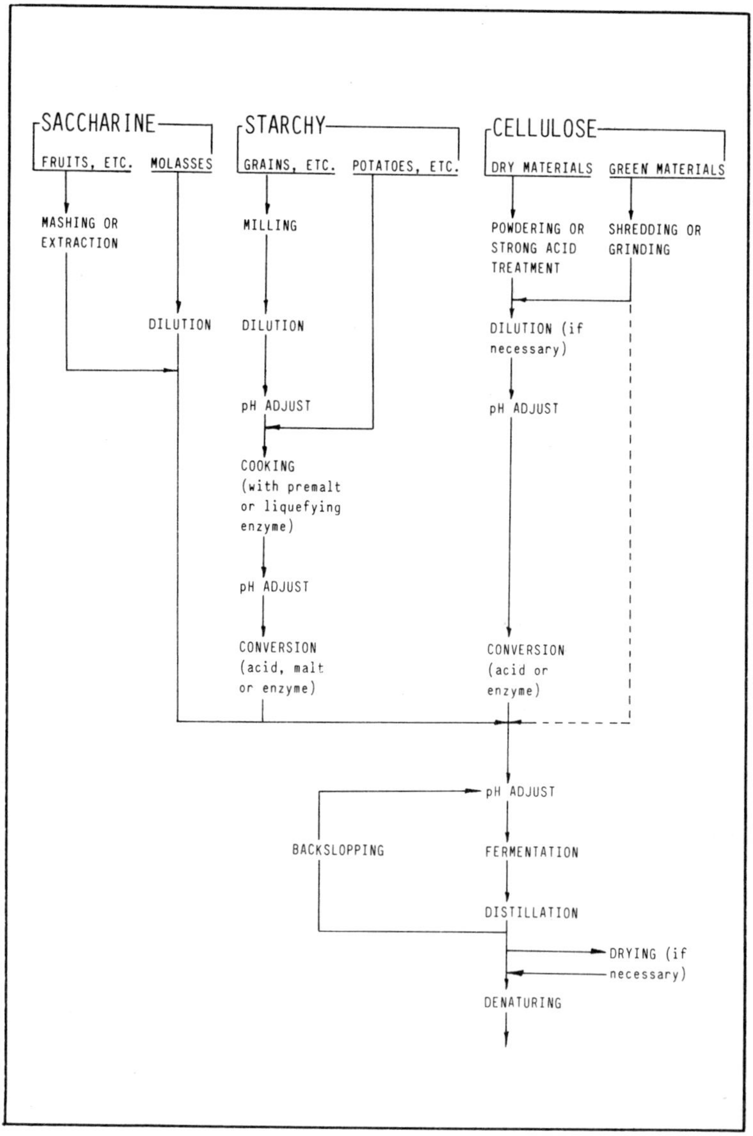

certain materials require less processing than others. Generally, small scale production is easiest (and most economical in terms of labor and energy consumption) from the saccharine materials. However, starchy materials usually produce the most alcohol on a weight/weight basis, and cellulose materials are the cheapest.

Manufacturing alcohol from saccharine feedstocks generally requires: (1) extraction or crushing to make the sugars available to the yeast enzymes during fermentation; (2) dilution, which is only required with certain materials; (3) fermentation; and (4) distillation. Starchy materials require the steps of: (1) milling to free the starchy material from, for example, grain kernels; (2) dilution; (3) cooking to dissolve and "gelatinize" the starch; and (4) conversion of the starch to fermentable sugars by malting, enzymes, or acid hydrolysis in addition to the steps of fermentation and distillation. Cellulose materials are similar to starchy materials in that they must be converted prior to fermentation.

PROCESS DESIGN

There are a great many variables in the manufacture of ethanol. Even materials from the same basic group can require radically different processing. The following chapters cover the individual manufacturing steps for processing each of the three main groups of feedstock. In addition, Chapter 10 contains "recipes" and individual processing requirements for specific materials from each of the three groups.

The reader is urged to study all of the information presented before attempting to choose a specific process for a material.

Chapter 5
PROCESSING STEPS COMMON TO ALL MATERIALS

DILUTION

Dilution is simply the addition of water to adjust the amount of sugar in the mash or (ultimately) the amount of alcohol in the beer. It is necessary because the yeast, used later in the fermentation process, can be killed by too great a concentration of alcohol. Also, during the mashing and conversion of starchy material, dilution is necessary to make the mash easier to stir and handle. The object of dilution is to end up with a beer as close to (but not more than) 10% alcohol when fermentation is complete. The optimum dilution, then, is a compromise between the highest alcohol concentration and the point where the particular yeast strain being used will be killed.

Optimum dilution requirements for each material are listed in Chapter 10. A rule of thumb for an unknown material, though, is that the final alcohol concentration will be about half the sugar content prior to fermentation. To determine the amount of fermentable sugar in a mash, it is best to have the material tested by a laboratory. If this is not possible, the sugar content can be estimated with a hydrometer. The use of hydrometers and tables for converting specific gravity readings to approximate sugar content are covered later in this chapter. It should be noted that any solution being tested with a hydrometer must be filtered to remove any undissolved solids. Otherwise the readings will be inaccurate. Sugar content of a solution can also be determined with the use of an optical instrument called a sugar refractometer. These devices, however, cost several hundred dollars.

Since the use of a hydrometer to measure

sugar content of a mash is, at best, an approximation, the amount of dilution can be "fine tuned" by measuring the alcohol content of the beer after fermentation. A hydrometer is used for this measurment also, but the readings are much more accurate. Naturally, if the alcohol content of the beer is less than the toleration level of the yeast you are using, the mash is overdiluted.

pH CONTROL

The pH is a measure of the acidity or alkalinity of an aqueous solution expressed on a scale of 1-14. Neutral is pH 7, pH 1-7 is acid, and pH-7-14 is alkaline. The pH is most conveniently measured with test papers that change color according to the pH of the solution being tested. These papers are available from swimming pool supply houses, garden shops, and laboratory supply companies.

Control of pH during the mashing and fermentation process is important for two reasons: The growth of harmful bacteria is retarded by acid solutions, and yeast will grow only in an (slightly) acid solution.

Most grain mashes have a naturally acid pH of between 5.4 and 5.6 after malting or conversion has been accomplished. Other materials, notably saccharine substances like molasses and fruit pressings, have a naturally alkaline pH and must be acidified prior to fermentation.

The principal bacterial contaminants in a distillery are those that form lactic acid. Although the production of fuel alcohol is not concerned with the taste of the product, any lactic acid formed subtracts from the yield of alcohol. The production of lactic acid and other contaminants should therefore be avoided as much as possible. The development of these micro-organisms is severely repressed at pH values under 5.0. Above 5.0 their growth is rapid. The optimum pH range then is 4.8 to 5.0. Anything below about 4.1 to 4.4 is detrimental to other (desirable) processes taking place during the mashing and fermentation. Consequently, the pH should be checked during the cooking and conversion. If it is much above 5.0, it should be reduced by the addition of acid.

The acid most commonly used is sulfuric, although any mineral acid is perfectly suitable. Hydrochloric (muriatic) acid, for example, is available from swimming pool suppliers. The acid should be added cautiously, the mash stirred, and the pH checked, because it is very important not to add too much. If you happen to add a little too much, the pH can be raised with sodium hydroxide (caustic soda) solution or with ordinary lime. But after a certain point, this is useless and the mash must be scrapped.

While adjustment during mashing is desirable, the proper pH during fermentation is absolutely essential. As soon as the pH in fermentation falls below about 4.1, the fermentation stops. If this occurs prior to complete conversion of the sugars, the yield will be low. On the other hand, yeast needs a slightly acid environment in order to grow. Consequently, the pH should be kept between

4.8 and 5.0 for optimum results.

There are two ways of adjusting pH. The first, as discussed, is the addition of acid. The second, and probably the best, is the addition of the naturally acid residues left from a previous distillation. These residues are called "stillage", and adding them to the mash is called "backslopping". Backslopping is discussed in more detail in the next section.

It should be stressed that the pH should be checked periodically during the fermentation as well as before. Certain fermentations will produce substances that alter the pH during the fermentation. Once the pH goes beyond the optimum range, attempts to salvage the process by adding acid or caustic soda do more harm than good. So keep a close watch and adjust before the pH goes out of range.

BACKSLOPPING

Backslopping, or the addition of still residues from the previous batch, has several advantages. First (as discussed in the previous section) is the adjustment of the pH to control bacteria growth. Second, the stillage provides nutrients that are needed by the yeast for rapid growth. The third reason is that the stillage provides a "buffering" action.

Grain mashes and starchy material generally provide enough nutrients for the growth of the yeast. Other materials, notably molasses and other saccharine materials, often do not. The addition of stillage can provide these nutrients where they are needed.

The buffer capacity of the mash is important. When an acid and a base are mixed together, they react violently to produce a salt. Buffering can be thought of as a barrier between the acid and the base that allows only limited contact and thus moderates the reaction. Grain mashes are generally well buffered between pH 5.0 to 6.0, poorly buffered between 4.4 and 5.0, and well buffered between 3.5 and 4.4. The addition of stillage aids in buffering the mash between 4.4 and 5.0. This provides stability and generally higher yields than mashes without stillage.

Different materials can tolerate differing amounts of backslopping. It is possible to have too much of a good thing, and too much baskslopping can be detrimental. The limits for various materials are discussed in Chapter 10.

CLEANLINESS

The cleaning of fermenting tubs, pipes, and the like is extremely important. If mash and fermentation residues are allowed to accumulate, bacterial contamination will be rampant and will greatly reduce alcohol yield.

Cleaning of the mashing and fermentation apparatus is usually done with steam in commercial operations. However, in a small plant, a thorough washing with disinfectant is usually adequate. Any disinfecting cleaner can be used, but, in the interest of economy, it is best to buy formaldehyde solution from a chemical supply house. For use it should be diluted 20:1 or more. Be advised that formaldehyde is a horribly foul smelling chemical that is intensely irritating to the skin, nose and eyes. The fumes also should not be inhaled. An alternate to formaldehyde is ammonia (ammonium hydroxide) solution, but the same cautions apply.

After disinfecting with formaldehyde or ammonia, the apparatus should be thoroughly washed out with clear water. It is best to clean equipment after every batch, but in some climates and at certain times of the year when the bacteria count is low, cleaning every second or third time might be all right. In any event, at the first sign of problems, a thorough cleaning is absolutely necessary.

HYDROMETERS

As illustrated in Figure 5-1, hydrometers are little floats with calibrated stems used to measure the specific gravity of a liquid. The most familiar example is the device used to check battery charge or anti-freeze protection.

Figure 5-1: HYDROMETER

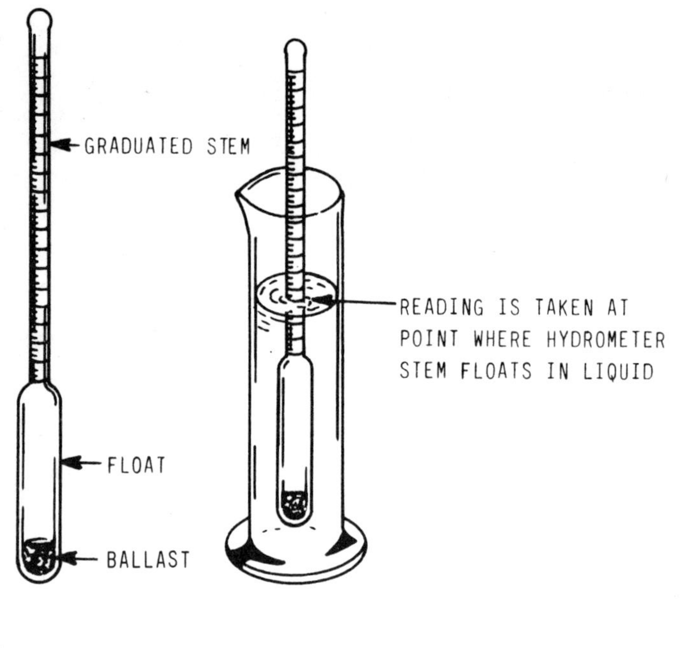

Hydrometers can be calibrated in a number of different scales, depending on their purpose. The most common calibration is for specific gravity. Water has a specific gravity of 1.000. Liquids lighter than water have specific gravities less than 1.000, and those heavier, greater than 1.000. The hydrometer can be used to measure the approximate dissolved solids in a mash or the concen-

Figure 5-2: SUGAR CONTENT vs SPECIFIC GRAVITY

This table represents percent dissolved sucrose in water as expressed in Specific Gravity @68° Farenheit.*

PERCENT SUGAR	SPECIFIC GRAVITY	PERCENT SUGAR	SPECIFIC GRAVITY	PERCENT SUGAR	SPECIFIC GRAVITY	PERCENT SUGAR	SPECIFIC GRAVITY
0	1.0000						
1	1.0039	11	1.0441	22	1.0918	42	1.1889
2	1.0078	12	1.0483	24	1.1009	44	1.1972
3	1.0117	13	1.0525	26	1.1102	46	1.2100
4	1.0156	14	1.0568	28	1.1195	48	1.2208
5	1.0196	15	1.0610	30	1.1290	50	1.2371
6	1.0236	16	1.0653	32	1.1386	52	1.2428
7	1.0277	17	1.0697	34	1.1484	54	1.2540
8	1.0317	18	1.0741	36	1.1583	56	1.2654
9	1.0358	19	1.0785	38	1.1683	58	1.2770
10	1.0400	20	1.0829	40	1.1785	60	1.2887

*Figures are accurate only for readings taken at 68° Farenheit. For readings at other temperatures, use correction table.

NOTE: This table will give values for pure sucrose in water. Because the liquid mash will contain other sugars, starches, and inert solids, the readings will be approximate and may not necessarily reflect the true sugar content. However, most readings will be within a few percentage points.

tration of alcohol before or after distillation. For measuring the solids dissolved in a mash, the hydrometer is calibrated in degrees "Balling". One degree on this scale is equal to about 1% dissolved solids. Other hydrometers can be purchased to show alcohol content in proof or percent. To obtain accurate measurements, a set of hydrometers, each covering a small range, is better than one hydrometer

Figure 5-3: ALCOHOL CONTENT vs SPECIFIC GRAVITY

Figures given as percent by volume @60° Farenheit, per U. S. Bureau of Standards.

ALCOHOL PERCENT	SPECIFIC GRAVITY	ALCOHOL PERCENT	SPECIFIC GRAVITY	ALCOHOL PERCENT	SPECIFIC GRAVITY	ALCOHOL PERCENT	SPECIFIC GRAVITY
0	1.0000						
1	.9985	26	.9698	51	.9323	76	.8746
2	.9970	27	.9687	52	.9303	77	.8720
3	.9956	28	.9676	53	.9283	78	.8693
4	.9942	29	.9665	54	.9263	79	.8666
5	.9928	30	.9653	55	.9242	80	.8638
6	.9915	31	.9642	56	.9221	81	.8610
7	.9902	32	.9630	57	.9200	82	.8528
8	.9890	33	.9617	58	.9178	83	.8553
9	.9878	34	.9604	59	.9156	84	.8524
10	.9866	35	.9591	60	.9134	85	.8494
11	.9854	36	.9577	61	.9112	86	.8464
12	.9843	37	.9563	62	.9089	87	.8434
13	.9832	38	.9548	63	.9066	88	.8402
14	.9821	39	.9533	64	.9043	89	.8371
15	.9810	40	.9518	65	.9020	90	.8338
16	.9800	41	.9502	66	.8997	91	.8305
17	.9789	42	.9486	67	.8973	92	.8270
18	.9779	43	.9469	68	.8949	93	.8235
19	.9769	44	.9452	69	.8924	94	.8198
20	.9760	45	.9435	70	.8900	95	.8160
21	.9750	46	.9417	71	.8875	96	.8121
22	.9740	47	.9399	72	.8850	97	.8079
23	.9729	48	.9381	73	.8824	98	.8036
24	.9719	49	.9362	74	.8799	99	.7989
25	.9708	50	.9343	75	.8773	100	.79389

*Figures are accurate only for readings taken at 60° Farenheit. For readings at other temperatures, use correction table.

covering a large range. Hydrometers can be purchased from any laboratory supply house. So that you do not have to purchase several sets, Figures 5-2 and 3 convert sugar and alcohol content to specific gravity.

Because liquids change density with changes

Figure 5-4: CORRECTION TABLE

Step 1:
Estimate the approximate percentage of water in the sample to determine correction factor listed below.

0 - 5%	= .00047	68%	= .00035	82%	= .00019
6 - 16%	= .00048	69%	= .00034	83%	= .00018
17 - 29%	= .00047	70%	= .00033	84%	= .00017
30 - 37%	= .00046	71%	= .00032	85%	= .00016
38 - 43%	= .00045	72%	= .00031	86%	= .00015
44 - 49%	= .00044	73%	= .00030	87%	= .00014
50 - 53%	= .00043	74%	= .00029	88%	= .00013
54 - 56%	= .00042	75%	= .00028	89%	= .00012
57 - 58%	= .00041	76%	= .00026	90 - 91%	= .00011
59 - 61%	= .00040	77%	= .00025	92 - 93%	= .00010
62%	= .00039	78%	= .00024	94 - 95%	= .00009
63 - 64%	= .00038	79%	= .00023	95 -100%	= .00008
65%	= .00037	80%	= .00022		
66 - 67%	= .00036	81%	= .00020		

Step 2:
Multiply the correction factor by the number of degrees greater or lesser than 60°

Step 3:
If temperature is greater than standard, add the correction to the specific gravity reading to obtain the temperature corrected reading. If temperature is less than standard, the correction should be subtracted from the reading.

in temperature, corrections must be made for readings taken at temperatures other than that for which the hydrometer is calibrated. More often than not, the temperature of the solution you are testing will not be the same as the hydrometer calibration. Figure 5-4 is a correction table for non-standard temperatures. Note that this table is very accurate for determining the correction for aqueous ethanol solutions and less accurate for sugar or dissolved solids concentration.

An alternate method of determining sugar content is the use of a sugar refractometer. These instruments are available from laboratory supply houses, but are expensive. However, the readings are very accurate. Specific gravity can also be determined by weighing exactly one liter of the liquid at the proper temperature. The weight in grams, divided by 1000 will be the specific gravity. Due to the difficulty of measuring exactly one liter of the solution being tested, the weighing method is usually not as accurate as the method using a hydrometer.

Chapter 6 PROCESSING STEPS SPECIFIC TO SACCHARINE MATERIALS

GENERAL DESCRIPTION

As stated earlier, saccharine materials require the least processing of any of the ethanol feedstocks. Referring to Figure 4-1, it can be seen that molasses and other sugar-containing syrups need only to be diluted and pH adjusted prior to fermentation. Other materials, such as grapes and other fruits, need to be either crushed or extracted to make the sugar readily accessible to the yeast enzymes.

In addition to extraction, the requirements of pH control, dilution, backslopping, and cleanliness as discussed in Chapter 5 also apply. Recipes for specific materials are contained in Chapter 10.

EXTRACTION

Prior to fermentation, saccharine materials, such as fruits, beets, etc., are usually put through an extraction process. This means that the sugar-containing juice is separated from the rest of the material. This is usually done in a press like those used for crushing grapes or making cider. Extraction, per se, is not absolutely necessary. The materials can also be simply crushed to expose the juices for the fermentation process. However, with most distillation equipment, the solids will have to be removed prior to going into the still.

There are certain problems and considerations associated with either extraction or crushing. Extraction in a press, at best, leaves an appreciable amount of fermentable material behind. Typically, only 75% of the sugars can be extracted from apples and about 80% from grapes. One technique that can be used with press extraction to increase the yield is to take the residue from the first pressing, soak it in a minimum amount of water to dissolve more of the sugar, and then press it again. However, this method creates additional dilution which lowers the alcohol content of the fermented mash (called "beer") requiring more energy and time in the distillation process. If a fruit juice contains, for example, 10% sugar, the final alcohol concentration going to the still will be about 5%. Any water used to wash additional sugar from the residue will further dilute this final concentration. The lower the alcohol concentration, the more water must be removed from the alcohol during distillation. However, in many instances, the greater total amount of alcohol gained justifies the additional dilution.

Crushing the material instead of extracting it in a press leaves all the sugar available for fermentation, although the material usually must be strained prior to distillation. Again, some of the valuable liquid will be retained in the residue and the only solution is to wash it with a little water. If you are using a simple pot still, such as described in this book, filtering the residue isn't absolutely necessary as long as the still pot is cleaned out after each run. In this case, the crushing method is superior.

Certain materials such as sweet corn stalks, sugar cane, and the like, require heavy hydraulic presses to effectively extract the juice. The alternate process here is to shred the material and then heat it with as little water as possible to dissolve out the sugar. Note that to obtain complete recovery of the sugar, the process described must be repeated several times. Again, a point is reached where dilution offsets the amount of sugar released and some compromise must be made. Note also that two extractions of one gallon each will dissolve more sugar than a single two gallon extraction.

Chapter 7
PROCESSING STEPS SPECIFIC TO STARCHY MATERIALS

PREPARATION OF STARCHY MATERIALS

Starchy materials fall into two main categories: (1) materials, such as grains, in which the starch is encased or protected by grain hulls; and (2) those materials, such as potatoes, where the starch is more readily available. Milling or grinding the material to expose the starch is necessary for the former group, but not the latter. Otherwise, all starchy materials require a certain amount of cooking and conversion of the starch to sugar prior to fermentation.

There are two basic methods of conversion. The first uses malt or an extract of the enzymes contained in malt and the second uses dilute acid in a process called "acid hydrolysis".

MILLING

Grains and similar starchy materials must be milled to expose the starches to the cooking, conversion and fermentation processes. The ideal is to grind the material as fine as possible without producing an excessive amount of flour. This is because fine (flour) particles are difficult to remove if the material must be filtered prior to distillation. Again, if you are using a simple pot still, the material need not be filtered and the presence of fine flour particles is not objectionable. Large amounts of flour can also make the mash too viscous (thick) and hard to handle. This is only objectionable if it must be pumped from container to container or otherwise handled. If you are doing everything in the same pot, the viscousness can often be tolerated. Otherwise, premalting, as discussed later, will solve the problem.

Almost any kind of grain-milling equipment can be used, or the grain can be milled by your local feed plant. Unfortunately, there is no alternate process, and if you are going to use grain as your feedstock, it will have to be milled.

COOKING

Cooking is necessary for all starchy materials. The object is first to dissolve all the water soluble starches and then, as much as possible, gelatinize them.

In commercial operations, cooking is almost always done with steam, under pressure, and usually in a continuous process. Water boils at 212° Farenheit at sea level and at a lower temperature as altitude increases. By using pressure cooking equipment, higher temperatures and shorter cooking times can be obtained. At 150 pounds pressure, for example, grain starches can be cooked in six minutes or less. Large scale pressure cooking equipment is expensive and, in this manual, the cooking times for various materials will be given for the "atmospheric" process where temperatures are in the 208-212° range. Cooking times for different materials are listed under the individual feedstocks in Chapter 10.

Because a lot of energy is needed to boil the water used in the cooking process, it is best to cook with as little water as possible. Then, after cooking, additional water may be added to dilute the mash to optimum concentration for fermentation. If the additional water is added at a time when it is desirable to cool the mash, for example after cooking and prior to conversion, cooling time is saved. Most grains can be cooked with as little as 15-20 gallons of water per bushel. Note that when cooking with minimum water, special attention must be given to stirring the mash. Otherwise lumping and burning may occur.

New methods of cooking are being developed that help to conserve energy. The most interesting is a method that combines milling and cooking into one operation without the use of water. The process uses heat generated by friction in the milling process to simultaneously cook the grain. It is all done in a specially designed grain mill.

CONVERSION

Conversion is the process of converting starch to fermentable sugars. It can be accomplished either by the use of malt, extracts of the enzymes contained in malt, or by the treatment of the starch (or cellulose) with acid in a process called "acid hydrolysis". Each method is discussed separately.

MALTING

Starch can be converted to fermentable sugars by the action of enzymes in malt. When the seeds of any cereal grain are moistened and allowed to sprout, certain enzymes (amylases) are produced which have the ability to convert starch to a form of fermentable sugar called maltose. All cereal grains produce these enzymes to a greater or lesser degree. However, barley produces by far the most and is usually the most economical to use.

You can either purchase dried, ground barley malt, or you can produce your own from the grain. However, undried or "green" malt, such as that you might produce yourself, will not keep unless dried. On a small scale, it is often better to use the commercial product.

In converting starch to sugar, malt enzymes exert two forms of chemical activity: liquefaction and saccharification. The intensities of these two activities depend on the temperature of the mash. The liquefying power is greatest at about 158° Farenheit. It begins to weaken at 175° and ceases at about 200°. The saccharifying (sugar making) power is strongest between 120-130° F. and is destroyed completely at 175°. Both of these actions are desirable. Therefore a compromise must be made. The conversion process is therefore usually begun after the cooked mash is allowed to cool to about 150° F. The material is held at this temperature for a certain length of time (depending on the material) and then allowed to cool to the optimum fermenting temperature.

The average malting recipe calls for the use of between $\frac{1}{2}$ to 1 pound of dried malt for each 10 pounds of grain. Again, specific recipes are covered later. The dried malt is usually mixed with warm water at a ratio of about $2\frac{1}{2}$ pounds per gallon to form a slurry. This slurry should be mixed about an hour ahead of time and added to the mash when it cools to the proper temperature.

Because barley malt is expensive, usually more expensive than the material it is used to convert, it is best to use as little as possible. The minimum amount can be determined after several trial conversions. To do this, make a trial malting using the amounts listed in Chapter 10. Then take a little of the converted mash and filter it through a cheesecloth or some similar material. Place a little of the filtered liquid in a white dish and add several drops of a solution composed of 5 grams potassium iodide and 5 grams of iodine crystals in one quart of (distilled) water. Any blue color produced indicates the presence of unconverted starch. Naturally, if the test indicates no blue color, the next trial should be run with less malt and vice-versa. The test solution can be compounded by your local druggist or the chemicals can be purchased from any laboratory supply house.

PREMALTING:

During the cooking process, the starch in the grain is gelatinized. When the mash is cooled, it may become too thick to be stirred and handled effectively during the malting operation. The technique of premalting cures this problem by taking advantage of the liquefying properties of malt prior to the conversion. To premalt, simply add about 10% of the total malt weight to the mash prior to cooking. This causes sufficient liquefaction to facilitate handling the mash during subsequent operations. It also helps to prevent thermal destruction of the malt enzymes later on and so reduces the production of undesirable by-products. After cooking, the remaining 90% of the malt slurry is added, and the conversion is continued as usual.

PREPARATION OF MALT

The following is a basic process for making malt, for those who prefer to prepare their own. Any grain can be used to make malt, but as stated before, barley is by far the best. However, if you are working with corn, for example, you can simply set aside about 20% of the grain, prepare a malt as described below, and use it in the same way you would use barley malt. The same is true of similar materials.

Malt is simply sprouted grain. The basic requirements for sprouting are moisture, warmth, and darkness. Grain can be sprouted in anything from a five-gallon plastic pail to a 55-gallon drum. The container should either have small holes poked in the bottom or, with larger containers, a valve protected by a screen or mesh that will allow water to drain but retain the material being sprouted.

Begin by soaking the grain until the kernels can be crushed between the fingers and the inside is soft. This takes about 8-12 hours for barley and considerably longer for corn. Then drain the water. Thereafter, sprinkle the grain several times a day with warm water. The object is to keep the grain moist but not wet. If too wet, the grain will rot. After a watering, the water will work its way down through the grain and out the holes (or valve) in the container. The sprouting will generate some heat. The optimum temperature for sprouting is about 80° F. but the most enzymes seem to be produced at about 60° F. When sprouting in large containers, be careful that the grain doesn't get too warm. If it does, it can be spread out on a concrete floor in a dark place and the sprouting continued. Small containers will not have the problem of too much heat.

Sprouting will take about 4 days. The malt is ready when the sprout is about $\frac{1}{2}$ inch long.

Prior to use, the malt will have to be crushed. This can be done in a mill or, on a small scale, a heavy duty garbage disposal can be adapted. It is also possible to use an ordinary blender or food processor.

Fresh, undried malt is called "green" malt. It must be used immediately or dried because it will rot if stored wet. It should be used in the same manner as dried malt, and it is not usually necessary to adjust the recipe to allow for the green malt's moisture content because the green malt is slightly more potent.

ENZYME CONVERSION

The enzymes contained in malt are available commercially from several manufacturers. The procedures for using them are very similar to malt conversion. In addition, the use of enzyme extracts is usually superior to malt. First of all, the enzyme extracts are usually cheaper. They are also specifically designed for the job at hand, and they generally produce more predictable results and higher yields.

The three basic types of commercially available enzymes are alpha, beta, and gluco amylases. Alpha amylases randomly split the starch molecules to produce a type of sugar called dextrose. Beta amylases act similarly to produce maltose. Together, these two enzymes can convert about 85% of the starch to fermentable sugar. Gluco amylases can reduce the remaining starches, and the use of all three can achieve almost total conversion of the starch.

The two principal manufacturers of enzymes suitable for starch conversion are Miles Laboratories and Nova Laboratories, as listed in the appendix. Enzymes are used in much the same manner as malt. However, because different enzymes require slightly different pH, times, and temperatures, it is best to follow the recommendations of the manufacturer.

A typical recipe for the use of Miles Laboratories "Taka-Therm" and "Diazyme L-100" for the conversion of corn is included in Chapter 10.

ACID HYDROLYSIS

Starch (and cellulose) may also be converted to fermentable sugars by the action of acid.

This process is relatively simple, but it requires acid proof equipment, high temperatures, and the handling of acid. For these reasons, it is not really recommended for small scale production.

Basically, dilute mineral acid (usually sulfuric) is added to the grain slurry prior to cooking at a concentration of 1-4% as calculated on a weight/weight basis. The mash is then cooked at a temperature of about 350° F. Cooking and conversion of the starch take place simultaneously. The mash is then immediately neutralized with calcium hydroxide (lime), or some other base, and fermented in the usual manner.

The high temperatures essential in this process are obtained by the use of pressure cooking. The steam pressure required is about 150 pounds per square inch. This, together with the necessity for acid proof equipment, make this process unsuitable for small scale use. However, it is an excellent process for large operations because cooking and fermentation times are short and the method is readily adaptable to continuous operation.

MASH COOLING

Malting is conducted at a temperature of about 145-150° F. As is discussed later, fermentation is commenced at an optimum temperature of 70-80° Farenheit. Between the two steps the mash must be cooled.

One of the biggest problems affecting alcohol yield is bacterial contamination of the mash either before or during fermentation. The chief protection against this is the pH or acidity control of the mashing and fermentation operations. However, even with perfect pH control bacterial infections can set in. This happens mostly during the cooling stage between mashing and fermentation.

If bacterial contamination becomes a problem, the only solution (other than the obvious need for cleanliness) is to shorten the cooling time as much as possible. The less time at the temperatures conducive to bacteria growth the better. Therefore, it might become necessary to make a cooling coil as illustrated in Chapter 13. The cooling coil is the best long term solution, but if the problem only occurs occasionally, as during the summer months, a plastic bag full of ice and suspended in the mash might do the trick. Just be sure the plastic bag doesn't leak and dilute your mash!

Chapter 8
PROCESSING STEPS SPECIFIC TO CELLULOSE MATERIALS

CELLULOSE CONVERSION

Cellulose feedstocks, which include a wide variety of material from corn stalks, wood, straw, and cotton, to old newspapers (paper) and trash, are potentially good sources of alcohol. If fully converted, for example, a ton of old newspapers would yield up to 70 gallons of alcohol. Cellulose materials are also extremely cheap and, often, free.

Cellulose is converted by either enzymes or acid hydrolysis. Nova Laboratories produces special enzymes called "Cellulast" and "Cellobiase 250L" for conversion of cellulose to fermentable glucose. Other manufacturers make similar products. The acid process involves either strong acid and relatively low temperatures, or weak acid and high temperatures. The strong acid process has the problem that the glucose is destroyed almost as fast as it is formed unless the contact time with the acid is very brief. The weak process requires acid proof pressure cooking equipment as described earlier. Again, for the obvious reasons, these methods are not recommended on a small scale.

The main problem with cellulose as an ethanol feedstock is getting at the cellulose itself. In a plant, cellulose is encased in a substance called "lignin" in much the same way that a steel reinforcing rod is encased in concrete. Lignin is the substance that gives wood its strength. To get at the cellulose, the lignin must be dissolved away. The paper industry uses substances like sulfur dioxide, calcium bisulfite, sodium sulfate, sodium sulfide, and sodium hydroxide (lye) to dissolve lignin. Concentrated mineral acid, mentioned earlier, also dissolves lignin. Unfortunately, as it is dissolving the lignin, the strong acid also converts and then destroys the glucose.

Commercial processes are being developed to process cellulose into alcohol with the use of strong acid without destroying the cellulose. However, the process is complicated and economically feasible only on a very large scale.

The only alternative to dissolving the lignin is to reduce the cellulose material to as fine a state as possible so that at least some of the cellulose may be recovered. This is done by powdering, grinding or pressing. The yield of cellulose is directly proportional to how finely the starting material is reduced.

Other cellulose materials are somewhat easier to process than those with high lignin content. Some forms of paper, like newspaper, are almost pure cellulose and are easily converted by either the enzyme or acid process.

Also, in order for a plant to produce cellulose, it must first produce glucose, which is the sugar we are trying to obtain. Therefore, plants that are processed while they are still wet and green have the advantage of having fermentable sugar already available. These materials can be simply fermented without conversion and considered as low-yield saccharine feedstocks.

Chapter 9
YEAST AND FERMENTATION

YEAST

Yeast is an organism belonging to the vegetable family. The yeast itself does not take a direct part in the fermentation process, but it secretes a complex of enzymes that act upon the sugar and convert it to alcohol and carbon dioxide gas.

The yeast used in alcoholic fermentation is a special strain bred to be tolerant to variations in pH and resistant to alcohol. In the past, distilleries bred and propagated their own yeast strains. The yeast was kept alive in cultures and grown in batches of ever-increasing size to be used in the fermenters. Keeping yeast alive and growing cultures is a tricky business that requires precise control of temperature, nutrients, and the like. However, a simplified method is described later. Fortunately, special active dry yeast is available. To use it, you merely add warm water to reactivate it and then add it to the mash in the fermenter. Two pounds is sufficient for 1000 gallons of mash. It is available from Universal Foods Corporation as listed in the appendix. This yeast should be rehydrated for 15 minutes prior to use at a temperature of 100-105° Farenheit, or it can be added dry to the fermentation tank prior to filling.

In a pinch, it is possible to use ordinary baker's yeast from your grocer's shelf. However, this yeast is not bred for alcohol tolerance, and you will probably not get the yields associated with the distiller's yeast.

YEAST PROPAGATION

It is possible to grow and propogate your own yeast cultures if you observe certain precautions. Above all, the conditions must be absolutely sterile. Ordinary boiling water does not kill all of the bacteria present. It is necessary to use a pressure cooker. Make a solution of (proportionately) one cup sugar, one cup flour and two quarts water. Place the solution in a pressure cooker and boil at elevated pressure for at least 45 minutes. Without opening the pressure cooker, cool the solution to about room temperature. Then open the container and add a cake of baker's or distiller's yeast. Close the container and keep it in the refrigerator. The yeast will slowly grow. Some carbon dioxide will be given off, so be sure to leave the vent open. If desired, the yeast slurry can be transferred to jars. Just be sure they are sterile and remember to poke a small hole in the lid to let the carbon dioxide escape.

To use the yeast culture, merely remove a teaspoon or so, place it in another (sterile) container, feed it some sugar and warm it to room temperature. When it becomes active, it is ready for the fermenter. If at any time your refrigerated culture goes bad (due to bacterial contamination) it must be thrown out and the procedure started again. Also, yeast cultures should not be frozen.

FERMENTATION

All that is necessary to begin fermentation is to mix the activated yeast and the cooled, pH-adjusted mash in the fermentation tank. Aside from the considerations of pH as discussed earlier, the most important thing during the fermentation is temperature control. When the fermentation begins, carbon dioxide gas will be given off. At the height of fermentation, the mash will literally "boil" from the carbon dioxide produced. The reaction also produces some heat. The optimum temperature for the fermentation process is between 70–85° F., and it is desirable not to let the temperature go much above 90–95° F. Cooling is readily done with the use of ice bags, as discussed earlier, or by the use of a cooling coil. A less desirable method of controlling temperature is to dilute the mash.

The actual time required to ferment a mash

varies with the material being fermented, the pH, temperature, and several other factors. It can take from one to four days. You will know that the fermentation is complete when the mash ceases bubbling and the yeast cake, which forms on top, sinks to the bottom. At this point, the fermented liquor is known as "beer" and it is ready to be distilled.

It is advantageous to distill the beer as soon as possible. Occasionally, if it is allowed to sit, it will turn to vinegar. Vinegar is alcohol that has been oxidized to acetic acid. Certain enzymes present after fermentation act as catalysts and allow any air present in the mash solution to react with the alcohol to form acetic acid. In fact, if you want to produce vinegar, all you have to do to start the reaction is to bubble air through the fermented mash. Once the vinegar reaction has set in, the mash is lost. There is no cure. The only prevention is to separate the beer from the mash sediment and distill it as soon after fermentation is complete as possible.

It is also advantageous to use a fermentation

lock as described in Chapter 13, to prevent alcohol vapors from escaping the fermenter. Otherwise, the CO_2 gas can carry with it a considerable amount of alcohol. Note that the small, glass fermentation locks available from wine-making supply houses are suitable, at most, for a 5-gallon container. Larger containers must have proportionately larger fermentation locks or a dangerous amount of pressure will build and the vessel could explode.

It is permissible to open the fermenter to check progress and take samples for pH analysis, etc. as long as care is taken not to introduce bacteria that could contaminate the mash.

FERMENTATION BY-PRODUCTS

The principle products of fermentation are alcohol, carbon dioxide, and fermentation residue. The alcohol is distilled from the beer and used as fuel. The carbon dioxide gas in large distilleries is usually compressed or made into dry ice. Another use for the gas would be to pipe it into a greenhouse. The plants will then use it in the photosynthesis cycle, removing the carbon and giving off oxygen. Lacking a use for the carbon dioxide, it can be simply vented into the air as it is totally non-polluting and non-toxic.

What will be left is a lot of water and solids. A portion of the water can be used for backslopping. The remaining solids contain proteins, vitamins, minerals, fats, and yeast cells. All of the nutrition value of the original feedstock, except the starch or sugar that has been turned into alcohol, survives intact. It may be fed to cattle, or if suitably processed it can be used for human consumption. However, in the wet state, it will keep for a maximum of 3-5 days depending on conditions. After this it will begin to rot. Therefore, for long term storage these residues (stillage) must be dried. This can be done by straining out the solids and spreading them in a thin layer to dry in the sun, by use of rotary grain dryers, or similar equipment.

NOTE OF CAUTION

Alcohol produced for human consumption is made under special conditions and purified to a high degree. Ethanol that is produced according to the procedures in this book will contain fusel oils (high boiling alcohols), aldehydes, and ketones. None of these chemicals affect fuel performance but, if ingested, could cause fatal poisoning at worst or a horrible hangover at best. In addition, if the distillation equipment used later on is not tinned copper or stainless steel, many toxic metal oxides can be introduced to the alcohol. Solder, for example, contains a lot of lead and can react to form poisonous lead oxides. So besides being illegal, drinking your fuel could be hazardous to your health!

Chapter 10
INDIVIDUAL RAW MATERIALS

INDIVIDUAL RAW MATERIALS

This chapter contains specific processing information and recipes for individual raw materials. If a material you are interested in using is not listed, you can usually approximate an appropriate process by using the information about a similar material. Note that feedstock materials are not consistent in the amount of fermentable materials, moisture content, and many other factors. The figures given here are averages. More specific information about particular materials can be obtained from your state agricultural service, or the material in question can be tested by an agricultural laboratory for a modest fee.

Remember, then, the following information is intended only as a guide.

SUGAR/STARCH CONTENT vs ALCOHOL

On the average, the amount of alcohol that can be produced from a given feedstock will be about half (on a weight/weight basis) of the convertible starch or sugar content. Ethanol weighs about 6.6 pounds per gallon. A ton of grapes, for example, with a 15% sugar content is capable (assuming 100% extraction) of producing about 150 pounds or 22.7 gallons of alcohol. Corn, with 66% convertible starch should produce 660 pounds or 100 gallons. Remember, this is only an approximation and actual yield depends on many interrelated factors.

SACCHARINE MATERIALS

The process of fermenting saccharine materials is relatively simple and straightforward. The steps involved are usually: (1) extracting or crushing, (2) pH adjustment through acid or backslopping, and (3) fermentation. Dilution is usually not necessary because the extracted juices often contain less than the 20% maximum of fermentable material. Exceptions to the above are the various types of molasses that do not require extraction, but usually require dilution.

FRUITS

The following are some fruits and their average sugar content: grapes, 15.0%; bananas, 13.8%; apples, 12,2%; pineapples, 11.7%; pears, 10.0%; peaches, 7.6%; oranges, 5.4%; prickly pear, 4.2%; watermelon, 2.5%; and tomatoes, 2.0%.

Allowing 75% extraction with apples, for example, the total fermentable material would be about 9% of the original weight. On this basis, a ton of apples would yield about ,13 gallons of alcohol. Assuming an 80% extraction with grapes, a ton should yield about 17 gallons. With watermelons and a 90% extraction, a ton would yield only about 3 or 3½ gallons. Clearly, some materials are better than others.

In all the above cases, the percentage of fermentable material in the extracted juice is low enough so that dilution is unnecessary and undesirable. To ferment these materials, the juice need only be adjusted to the proper pH (between 4.8-5.0) and the yeast added at the usual rate of 2 pounds per 1000 gallons of mash. To provide proper nutrients to the yeast, backslopping of about 20-25% by volume is desirable.

Also, all of the above materials may be simply crushed or pulped instead of extracted in a press. This way the total sugar content is available for fermentation. If you are using simple batch distillation equipment that does not require the beer to be strained, this method is recommended.

MOLASSES

Beet or cane molasses is the residue from the manufacture of sugar. These materials, if available, are excellent sources of alcohol. They contain 50-55% fermentable sugar, and a ton should yield between 70-80 gallons of alcohol.

Molasses with a sugar content above 15-20% will need to be diluted. Since most molasses is low in the nutrients necessary for proper yeast growth, backslopping is of particular advantage. Up to 50% stillage (by volume) may be used. Also, most molasses is naturally alkaline, and acid will be needed in addition to the stillage to obtain the proper pH value.

CANE SORGHUM

Cane sorghum is a good alcohol source because it is easily grown and averages about 14% fermentable sugar content. The main drawback to using this material is that the extraction requires heavy-duty shredding and pressing equipment. An alternate process is to shred the stalks as much as possible and dissolve the sugar by heating (not quite to a boil) with a minimum amount of water. The process must be repeated several times to retrieve most of the sugar. Note that in this type of process, two extractions of one gallon each are better than one extraction of two gallons.

A conservative 65% extraction should yield about 13-14 gallons of alcohol per ton. Acidification to proper pH is necessary and backslopping to about 25% can be tolerated.

SUGAR BEETS

Sugar beets are an excellent material for ethanol production. They contain about 15% sugar, 82% water, and the rest in various solids. The juice can be extracted in a press, or the beets can be crushed and fermented as decribed in the section on fruits. Because the beets contain a certain amount of starch, the addition of small quantities of malt (1-2% by weight) or enzyme will greatly improve the alcohol yield. Adjustment of pH is, of course, necessary, and backslopping in the 20-25% range is desirable. A ton of beets should produce 20-25 gallons or more of alcohol.

SUGAR CORN WASTES

Stalks from sugar corn contain 7-15% sugar and should be considered as an alcohol source if they are available. The stalks need to be shredded and extracted in a manner similar to sugar cane or sorghum stalks. A relatively efficient operation should yield 8-18 gallons of alcohol per ton of material. Again, back-slopping to 20-25% and acidification are necessary.

STARCHY MATERIALS

Starchy materials generally require milling, cooking, and conversion prior to fermentation. Exceptions are materials, such as potatoes and sweet potatoes, that do not require milling, and materials, such as artichokes, that do not require conversion. Relatively high alcohol yields often offset the necessary additional manufacturing steps, and most starchy materials are good alcohol sources.

GRAINS

Grains must be milled, diluted, cooked, and converted prior to fermentation. However, they contain large amounts of potentially fermentable material. The average content of convertible starch and sugar in some typical grains are: barley, 50%; maize (indian corn), 66%; oats, 50%; rye, 59%; sorghum seed, 67%; and wheat, 65%. Alcohol yield per ton is dependent on how completely the starches are converted to fermentable sugar, but should be between 70-100 gallons.

After milling, the grain must be diluted prior to cooking and fermentation. The average dilution is between 56-64 gallons per 100 pounds of grain, depending on moisture and starch content. The method of cooking with minimal water and adding the balance prior to conversion, as described previously, has the dual advantage of reducing the energy needed for cooking and shortening the cooling time. Premalting with malt or enzymes is generally desirable.

Cooking is accomplished by heating the diluted and premalted mash to a slow boil and holding at this temperature for 30-60 minutes. Generally, the mash is sufficiently cooked when it is soft and mushy. The mash is then cooled to 145-150° and the malt slurry is added. The malt slurry consists of about $2\frac{1}{2}$

pounds of dried or green malt per gallon of water as described in the section on malting. On a weight/weight basis, corn or wheat will require about 8-10 pounds of malt per 100 pounds of grain. Rye will require about 10-12 pounds of malt for the same 100 pounds of grain. Other grains will fall somewhere in between. The malt slurry is stirred constantly during conversion. For wheat, the conversion will be complete in 5-15 minutes. Corn will require about 30 minutes, and rye between 30-60 minutes. The actual time, as well as the minimum amount of malt necessary, can be determined through trial mashes and the starch test as described in the section on malting.

When the conversion is complete, the mash is cooled to 70-75° F. and yeast slurry is added. Note that most grain mashes have an acceptably low pH and often do not need much adjustment. Backslopping should be limited to 20-25%.

The following is the general procedure for converting corn with Miles Laboratories enzymes. The procedure for other materials and other enzymes will differ slightly, and the manufacturer's recommendation should be followed.

After milling, the grain is partially diluted (slurried) at a ratio of 35 gallons of water per 100 pounds of grain. The pH is adjusted above 5.5 with an optimum range of 6.0 to 6.5. "Premalting" or liquefaction, is accomplished by the addition of 0.3 ounces of Taka-Therm enzyme.

The mash is then slowly heated. Gelatinization will begin at about 150° F. and the mash will rapidly thicken. Constant stirring is necessary at this point. At about 160° the liquefying action of the enzyme will begin. Heating may be more rapid after the liquefying action of the enzyme begins to take effect. After the mash reaches 200-212° an additional 1.3 ounces of Taka-Therm enzyme is added.

After the mash has been held at a slow boil for 20-30 minutes, an additional 33 gallons of water is added to complete dilution and cool the mash.

When the mash has cooled to 135-140° the pH is adjusted to 4.2 with acid and Diazyme L-100 enzyme is added at a ratio of 4 ounces per 100 pounds of grain. This enzyme completes the conversion in about 30 minutes and, after cooling to 70-80°, the mash is fermented in the usual manner.

JERUSALEM ARTICHOKES

Jerusalem artichokes deserve special mention as a source of alcohol because they contain between 16-18% fermentable material. In addition, the starches present can be converted without the use of malt or enzymes if cooked for a sufficient length of time. A ton should yield about 25 gallons of alcohol. To prepare artichokes for fermentation, they should be crushed to a pulp and cooked for 2-3 hours. If the starch test (described in Chapter 7) indicates that some unconverted starch is still present, conversion with small amounts of either malt or enzyme might be needed. Shorter cooking times are possible if a greater amount of malt or enzyme is used. For example, a 30 minute cooking time should be sufficient with a conversion using 3-6% malt or the equivalent amount of enzyme. Dilution is not necessary because the root usually contains 79-80% water. After cooking, the pH is adjusted and fermentation commenced in the usual manner.

POTATOES

Potatoes contain between 15-18% fermentable material and are a traditional source of alcohol. On the average, a ton of potatoes will yield about 22-25 gallons of alcohol. Damaged or sprouted potatoes are not objectionable, and the use of sprouted potatoes will reduce the amount of malt or enzyme required for conversion.

Commercially, potatoes are usually cooked with steam, under pressure. An acceptable alternate method is as follows: The potatoes should be shredded or cut up and placed in the cooker with as little water as possible; cover the cooker and steam until the potatoes are reduced to a soft mass. Premalting to reduce viscousness is a definite advantage. After cooking, the mash is cooled to the conversion temperature. Usually only 3-4 pounds of malt per 100 pounds of potatoes are all that is required. The mash must be constantly stirred during conversion, which will take about 15-20 minutes.

Because cooking and conversion times will vary, depending on starch content and the like, test mashings and the use of the starch test is recommended. Once converted, the pH should be checked and the mash fermented in the usual manner.

For specific procedures for the use of enzymes to convert potatoes, consult the manufacturer. Otherwise, about half the amounts listed in the corn recipe should be sufficient.

SWEET POTATOES

Sweet potatoes average about 22% starch and 5-6% sugar for a total of 27-28% fermentable material. A ton should yield up to 40 gallons of alcohol. Sweet potatoes are cooked and converted in a manner similar to potatoes with the exception that they contain only about 66% water and some dilution is necessary.

CELLULOSE MATERIALS

The following "recipe" for the conversion of cellulose is based on the use of two enzymes available from Novo Laboratories. "Celluclast" enzyme is produced from a variety of the Trichoderma viride fungus and is active in breaking cellulose into celloboise and glucose. The former is not a fermentable sugar, therefore, a second enzyme, "Cellobiase" is used in conjunction with Celluclast to convert the cellobiose. Together the two enzymes have the ability to convert cellulose to sugar with near 100% efficiency.

However, in order for the enzymes to work, the cellulose must be accessible. Any cellulose material should be shredded, ground, or otherwise reduced to as fine a state as possible.

After shredding, or whatever, the material is mixed with as little water as possible to make a thick, soupy mass. The pH is adjusted to between 4.5–6.0, and the enzymes are added.

It will be impossible to determine the exact amount of "accessible" cellulose, and the amount of enzyme needed must be estimated. Generally, dry cellulose materials such as wood, straw, corn cobs, etc. will have the lowest yields. This is because the cellulose is encased in lignin, and the amount that is ultimately accessible to the enzymes is proportional to how finely the material is divided. Materials such as grass clippings and all moist, green cellulose containing materials will have the next highest yields. This is partly because the lignin content is lower and partly because some fermentable glucose is already present. The highest yields will come from materials such as paper and cotton that are almost pure cellulose.

A trial amount of enzyme for all of the above materials would be about 2% Celluclast and 0.2% Cellobiase on a weight/weight ratio to available cellulose. Thus, if wood chips were estimated to have 5% by weight available cellulose, about 0.1% of the first enzyme is needed and 0.01% of the second. This would work out to 32 ounces of Celluclast and 3.2 ounces of Cellobiase per ton of wood. Newspapers, on the other hand might have 50-80% available cellulose and the amount of enzyme needed would be greater.

Optimum temperature for the enzyme reaction is 140° F. The mash should be held at this temperature for about 16 hours. The temperature should then be reduced to $80\text{-}90^{o}$ and fermentation commenced in the usual manner. Prior to adding the yeast, the pH should be checked and adjusted to the optimal range for the yeast strain.

It is suggested that trial conversions and fermentations be made to determine the minimum amount of enzyme needed to produce maximum yield.

A simplified "recipe" for green cellulose material of almost any kind is shredding followed by fermentation. As noted earlier, plants first produce glucose (a fermentable sugar) and then convert the glucose to cellulose. Yields will be based entirely on the amount of glucose present as cellulose is not converted by this method.

MULTIPLE ENZYME TREATMENT

All materials used in the production of ethanol will contain some cellulose. Therefore, it might be worthwhile to experiment with small amounts of cellulose enzymes in conjunction with the other processes. Saccharine materials might benefit from a separate cellulose conversion step. Starchy materials could have the cellulose enzymes added during conversion in addition to the starch enzymes. Depending on the amount of available cellulose, this procedure could dramatically increase yields.

Chapter 11
DISTILLATION

DISTILLATION THEORY

The object of distillation is the separation of the alcohol from the other ingredients in the beer, mostly water. In making fuel alcohol, it is necessary to get all of the alcohol and water separated if the alcohol is going to be mixed with gasoline, and most of the alcohol and water separated if the alcohol is going to be burned in a converted engine. As will be seen, the purer the alcohol, the harder it is to make.

The separation of the alcohol and water by distillation is made possible by the fact that alcohol boils at about 173° F. and water at

Figure 11-1: SIMPLE DISTILLATION APPARATUS

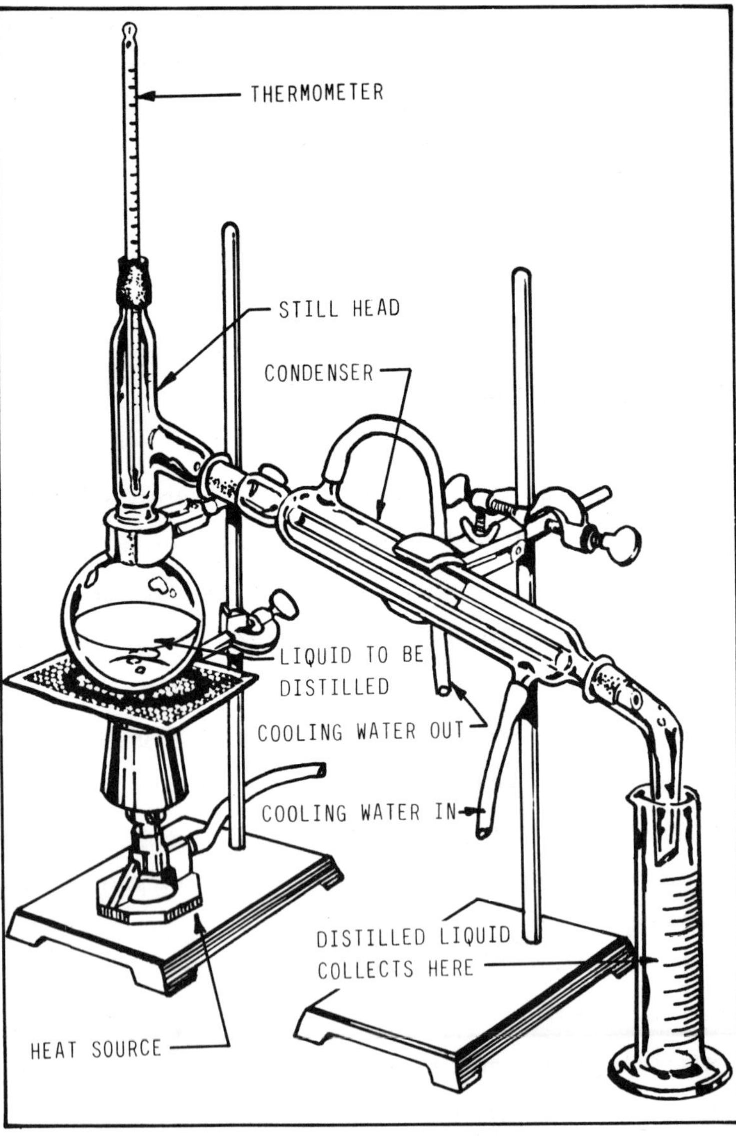

212° F. When the mixture of water and alcohol is boiled, vapors with a greater concentration of alcohol will be formed and liquid with a lesser concentration of alcohol will remain behind. However, because water and alcohol do not form what is called an "ideal" mixture, the separation cannot be done in one clean step.

Figure 11-1 illustrates a simple distillation apparatus using laboratory-type equipment. Note that the equipment consists basically of a container for the liquid to be distilled (still pot), a heat source, and a condenser to turn the distilled vapors back into liquid form. The thermometer is necessary to monitor the temperature of the vapors.

In Figure 11-2, the heavy solid curve represents the composition of the liquid phase of water/ethanol mixtures plotted against the temperature at which the mixture boils. The dotted curve represents the vapor phase. Using the apparatus illustrated, and starting with a concentration of 8% alcohol in water, the liquid will boil at about 200° F. Reading

Figure 11-2: BOILING POINT COMPOSITION for LIQUID and VAPOR PHASES

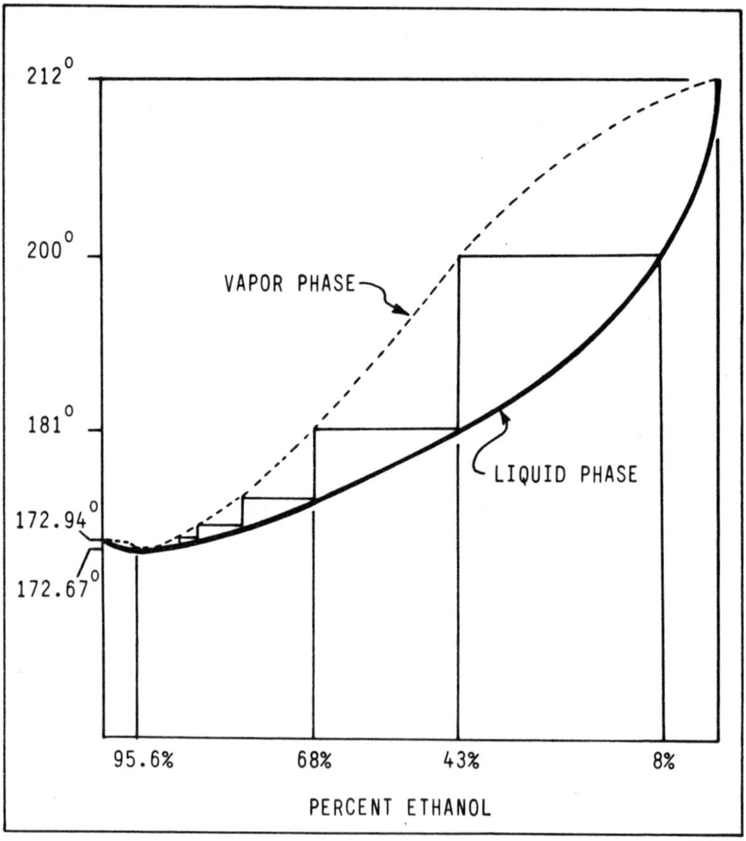

across, the vapors will contain about 43% alcohol. Clearly, for fuel purposes, a purer product is needed. To this end, we must redistill the condensed vapors from the first distillation which contain 43% alcohol and 57% water. This mixture will boil at about 181° F.

and the vapors will contain about 68% alcohol. Each time the condensed vapors are redistilled, they will be slightly purer, but many separate distillations are needed to produce relatively pure alcohol. Fortunately, a type of distillation apparatus, called a reflux (or rectifying column), in effect, performs simultaneous distillations and will be described later.

However, with any distillation equipment, even the most elaborate, the purest alcohol that can be produced is 95.6%. The remaining 4.4% water is impossible to remove because at this ratio, water and alcohol form a constant boiling mixture (called an "azetrope") whose boiling point is a fraction of a degree below that of pure alcohol, and separation by ordinary distillation is impossible. Special techniques that can remove this residual water are outlined later in Chapter 12.

THE REFLUX COLUMN

Figure 11-3 illustrates a reflux column installed on the simple apparatus described in

Figure 11-3: REFLUX DISTILLING APPARATUS

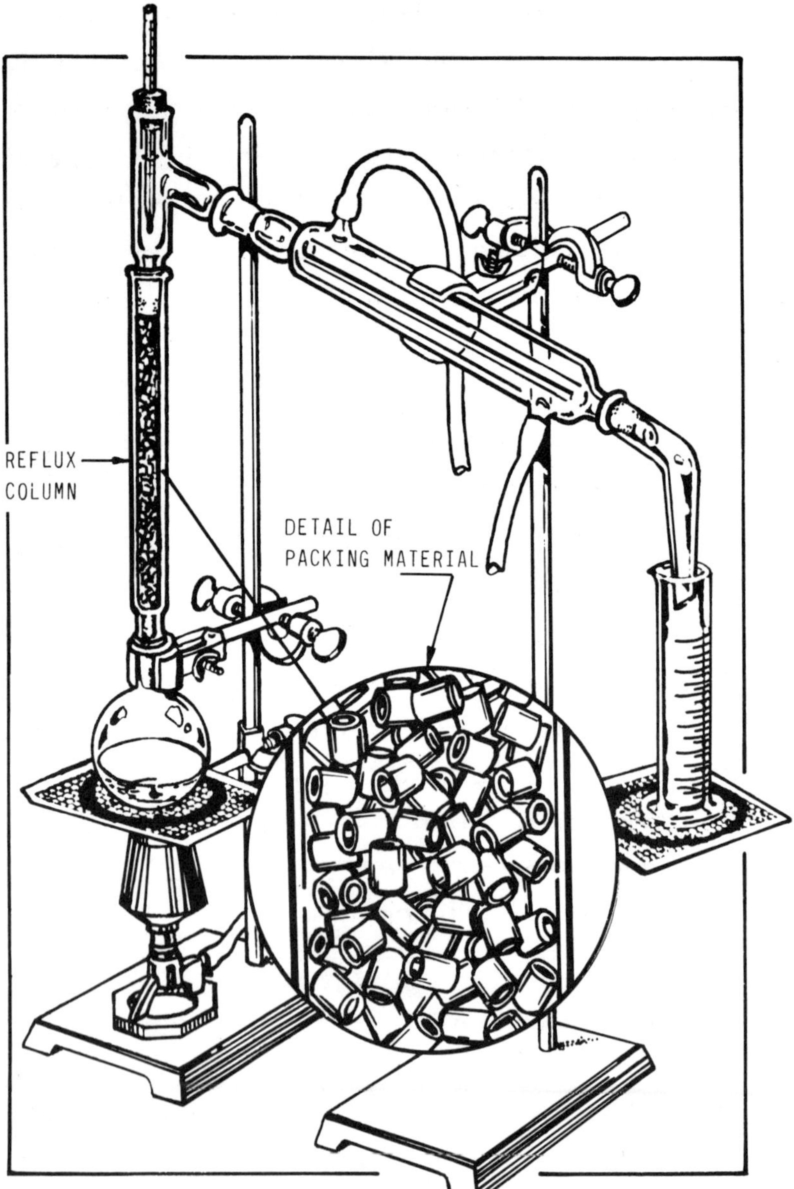

Figure 11-1. In this laboratory version, the reflux column consists merely of a glass tube filled with packing material. In this case, the packing is short lengths of small-diameter glass tubing. The purpose of the packing is to provide as large an internal surface area as possible.

As the vapors from the still pot ascend through the column, they condense on the packing material and drip downward. Additional ascending vapors contact the descending liquid (called "reflux") and revaporize it. Thus, as the vapors slowly work their way up the column, they become richer and richer in alcohol until, when they reach the top, they are relatively pure. Meanwhile, the descending liquid is stripped of its alcohol. The overall effect is that many "distillations" are performed simultaneously and the liquid in the still pot is stripped of its alcohol in one continuous operation.

Reflux columns can be constructed to operate on either a batch or continuous basis and are described in Chapter 14.

Chapter 12
DRYING THE ALCOHOL

GENERAL DESCRIPTION

As stated earlier, if the alcohol you produce is going to be burned in an engine modified to burn "pure" alcohol, water content of 5, 10 or even 15% is not objectionable. However, if the alcohol is going to be mixed with gasoline, the water will either have to be removed or "neutralized".

Drying the alcohol is a separate manufacturing step requiring additional labor, energy and/or expense. It would be wise to consider all factors before deciding whether this procedure is necessary.

ABSORPTION METHODS

If you are able to obtain alcohol from your still that is close to 190 proof (95%), the easiest method of producing "gasohol" is to mix sufficient benzene (benzol) with the alcohol to keep the water from separating. Usually, benzene in a ratio of about 2-3 times the amount of water in the alcohol is sufficient. For example, to prepare 100 gallons of 10% gasohol, mix 10 gallons of 95% alcohol with 2 gallons of benzene. Then add 88 gallons of gasoline. Note that water percentages greater than 5% require the addition of much larger amounts of benzene. It is suggested that experiments on a small scale be done before attempting to mix large quantities of gasohol. Prepare a mixture according to the above proportions and see if it separates. You may want to put a sample in the refrigerator to test cold weather tolerance. At any rate, if the mixture separates, more benzene is needed.

A second method takes advantage of the fact that water will dissolve in most salts, but ethanol will not. Therefore, water can be removed (although not entirely) by filtering the alcohol/water through dry salt. Almost any "hydroscopic" (water absorbing) material such as calcium salt, various sulphates, phosphates and similar materials will work. However, common rocksalt, such as used in water softeners is cheap and available. An apparatus such as described in Chapter 7 for sprouting malt can be used. Fill the drum or container with rocksalt. The alcohol/water is poured in at the top and filters down through the salt. Relatively water-free alcohol is collected through holes or a valve at the bottom of the container. Remember that the salt must be dry. After absorbing water from a certain amount of alcohol, the salt must be re-dried either in an oven or by spreading it out in the sun.

A good system might be to use both of the methods described above. First, most of the water in the alcohol is removed by the salt method, and then the blend is prepared with benzene. The more water that can be removed from the alcohol, the less benzene will be needed. Benzene will have to be purchased (the cheapest technical grade is fine) and can be used only once. The rocksalt can be dried and used many times.

DRYING WITH LIME

The oldest method of drying alcohol is dehydration with lime. This process is still used on a laboratory scale. It is similar to the salt absorption method just described except that, with this method, water is removed by a chemical reaction. Ordinary lime (calcium oxide, formula CaO) reacts with water to form calcium hydroxide (formula $Ca(OH)_2$). The process is simple. The water-containing alcohol is mixed with lime at a ratio of about 35 pounds (or more) of lime for each gallon of water to be removed (as determined with a hydrometer) and allowed to "slake" for 12-24 hours with occasional stirring. The lime reacts with the water to form calcium hydroxide. The calcium hydroxide is insoluble in the alcohol and so the relatively pure (99.5%) alcohol goes to the top of the container and the calcium hydroxide settles to the bottom.

The usual method of separating the lime and calcium hydroxide from the alcohol is by distillation. Alternately, but less desirable, the alcohol can be carefully drawn off (decanted) and filtered to remove any suspended particles that give it a milky appearance.

An apparatus based on a 55-gallon drum (similar to that in Figure 13-1) can be built. A still head with thermometer (no reflux column is needed) and condenser should be added to allow simple distillation. Also, a small gate valve located 6-8 inches from the bottom (above the level of the lime) will allow the alcohol to be decanted, if desired.

After slaking in the apparatus, the alcohol should be distilled off through the simple still head and condenser. During distillation, the temperature should remain exactly at the boiling point of pure alconol. Remaining after the pure alcohol has been distilled or decanted is wet calcium hydroxide and lime.

Some alcohol will also be trapped in the residue. To recover it, continue the distillation. The still head temperature will rise above 173° indicating that water is coming over with the remaining alcohol. When the still head reaches $208–212^{o}$ F. all of the alcohol has been removed. The water/alcohol distillate should be added to the beer for the next run in the reflux distillation apparatus.

The calcium hydroxide may be converted back into calcium oxide and re-used. However, the temperatures required are quite high unless a vacuum drying oven is used. Since lime is relatively cheap, this process is not recommended.

AZETROPIC METHODS

Commercial alcohol distilleries usually use an "azetropic" drying system. The fact that water and alcohol form a binary (two-part) azetrope, or constant boiling mixture, was described earlier. If a hydrocarbon solvent such as benzene, trichlorethylene, or even gasoline is added to the water/alcohol mixture, a ternary (three part) azetrope is formed.

Typically, trichlorethylene is mixed as a vapor with the 190 proof alcohol/water vapors coming from the rectifying column. These vapors are passed to another column. The water/trichlorethylene/alcohol azetrope boils at about 153° F. which is lower than the boiling point of pure alcohol. Therefore, the water, trichlorethylene, and some alcohol will vaporize first and exit the top of the column while pure anhydrous alcohol will exit the bottom. The water and trichlorethylene will separate into two layers upon standing in much the same way that water and gasoline will separate. Because of this separation, the trichlorethylene can be recovered and reused. The water (recovered from the trichlorethylene) contains some alcohol and is passed back into the main apparatus and redistilled. Systems using benzene and gasoline work in exactly the same way except that the boiling points are slightly different. If gasoline is used as the solvent, gasohol is produced automatically!

It is possible to mix benzene or gasoline in a reflux batch still, as described in Chapter 14, and boil off the water/solvent through the rectifying column leaving pure, dry alcohol behind. However, in doing this you would be boiling a highly flammable mixture of solvent and alcohol. This could be extremely dangerous, especially if an open fire were used to heat the still. Please do not attempt this procedure unless you are absolutely sure you know what you are doing and the equipment is properly designed.

Chapter 13
MASHING AND FERMENTATION EQUIPMENT

GENERAL DISCUSSION

Milling, extracting, cooking, conversion, and fermentation equipment can be designed to operate on either a batch or continuous basis. Batch equipment can be built in almost any size. However, it does not lend itself to complete automation and, thus, is more labor intensive. Continuous equipment can be fully automated, but it is expensive to engineer and construct.

An example of continuous cooking is a "jet" cooker. The mash slurry is injected into a pipe along with steam at about 150 pounds pressure. It is quickly cooked (usually in a matter of minutes) and emerges from the other end of the pipe ready for cooling and conversion. If the mash slurry is acidified (as described in Chapter 7), cooking and conversion take place simultaneously, and the mash emerges from the cooker ready for fermentation.

Slightly less complicated is the use of high pressure steam for batch cooking. Here a pressure vessel is used instead of the jet cooker. Although the operation is not continuous, cooking times are very short and simultaneous cooking and conversion are possible with the acid hydrolysis process.

Fermenters can also be run on a continuous basis. They are simply connected in series so that the mash flows from one container to the next. Continuous fermentation is best suited to continuous cooking and conversion systems where all the variables are controlled. It is very important that the feed going into a continuous fermentation system be absolutely sterile, otherwise the whole system can quickly become contaminated. Cooking with high pressure steam achieves the necessary degree of sterilization, most other processes do not.

Continuous systems, in general, are best suited to processing homogenous feedstocks of consistent quality. Batch operation is more advantageous when there are variables in feedstock quality that require adjustment of cooking, conversion and fermentation processes. Continuous equipment is also extremely expensive, whereas batch systems are less complicated and easier to construct. The following pages discuss the construction of batch equipment of different sizes and degrees of sophistication.

BATCH COOKING AND MASHING EQUIPMENT

Cooking and mashing equipment can be as simple or as elaborate as you wish. Figure 13-1 shows a 55-gallon drum with a large gate valve installed at the bottom. Figure 13-2 shows the same 55-gallon drum mounted over a simple firebox. This size apparatus is capable of processing about a bushel (56 pounds) of

Figure 13-1: SMALL APPARATUS

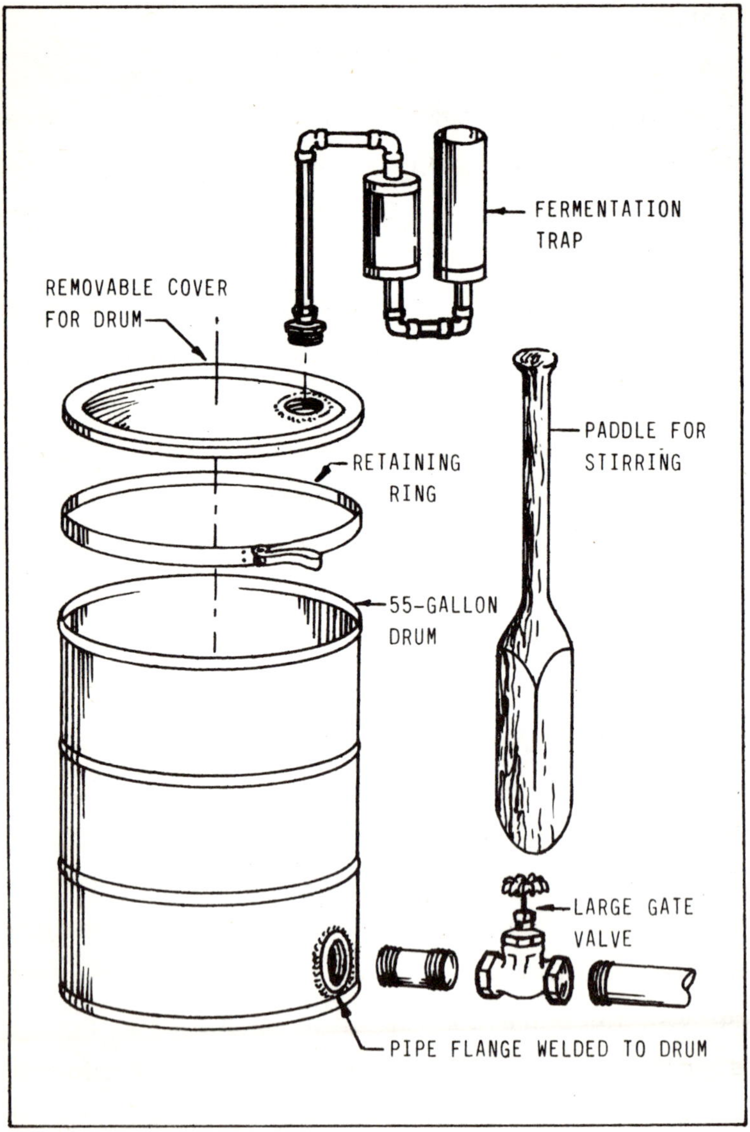

Figure 13-2: APPARATUS ADAPTED for COOKING

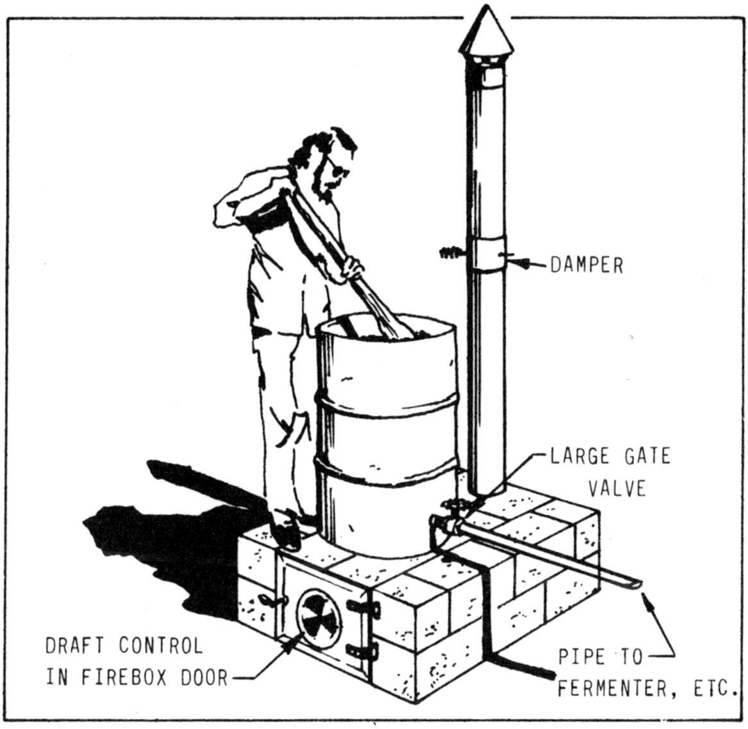

grain on each run. After cooking and conversion, the gasket-sealed top and fermentation trap would allow the same container to be used for fermentation. After fermentation is complete, the addition of a simple reflux column as described in Chapter 14 would allow the same drum to be used for distillation. Used in this manner, a single 55-gallon apparatus would produce about 3 gallons of alcohol every 3 or 4 days. By adding 2 or 3 separate fermenters of 50-gallon capacity and cooking and distilling one batch per day, alcohol production could be increased.

Figure 13-3 shows a large apparatus with certain improvements. Here a 3-4 inch centrifugal pump is used to stir the mash. Note that when using a pump for stirring, the larger the diameter of the pump, the less chance of clogging. However, even with large diameter pumps, full dilution at the cooking stage is sometimes necessary to insure adequate agitation. A 3000 gallon apparatus could prepare up to 200 gallons of alcohol every 3-4 days. The addition of separate fermenters would allow production of the same amount each day.

Tanks can also be rigged with paddle-type stirrers either vertically or horizontally. Also, a "jet" mixer, which looks a little like an outboard motor, is easily clamped to the side of an open tank.

Aside from the simple "tank over a fire"

Figure 13-3: LARGE APPARATUS

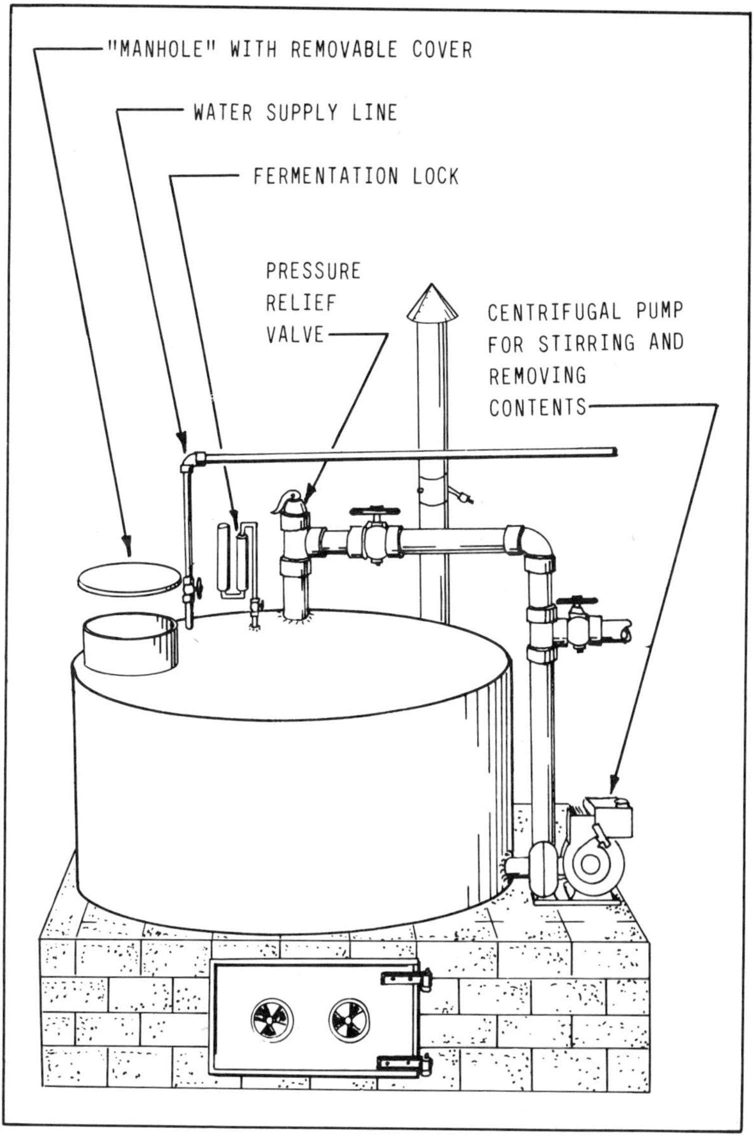

apparatus shown here, mash can be cooked by any other convenient source of heat. Perhaps the best would be the use of direct steam. This method requires only the absolute minimum initial dilution, or no dilution at all. Steam is simply piped into the bottom of the cooker and gelatinizes the mash. This method is particularly advantageous with potatoes where the 80% moisture content makes dilution undesirable.

One final note is that whatever source of energy you use for cooking, the apparatus should be insulated as much as possible. The better insulated the cooker, the less heat will be radiated away and wasted.

FERMENTATION EQUIPMENT

A fermenter can be made from just about any suitable tank. On a small scale, heavy duty 44-gallon plastic trash cans with sealable lids and fitted with a fermentation lock work very well. On a larger scale, plastic storage tanks, available up to 6000 gallon capacity are also excellent. The plastic has the

Figure 13-4: COOLING COIL

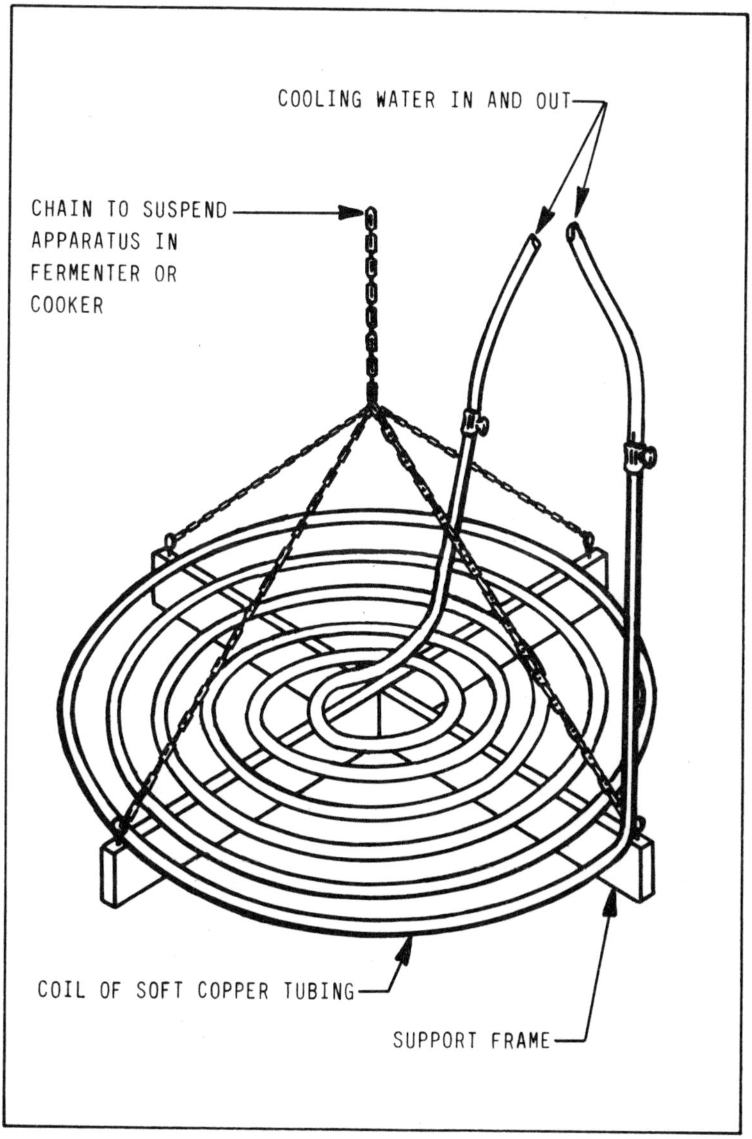

advantage of being rust-free and easily cleaned. However, if kept clean, ordinary steel tanks or drums should pose no problem.

The other requirements for a fermenter are some sort of agitation and a means of cooling. Agitation keeps fresh nutrients available to the yeast cells. Agitation can be achieved with either pumps or mechanical stirrers as described in the previous section. Cooling is necessary, especially in large fermenters, to keep temperatures within the prescribed limits. Figure 13-4 shows an easily constructed cooling coil. The unit shown is a coil of soft copper pipe on a suitable support frame that can be moved, for example, from fermenter to mash cooker, etc. Alternately, an ordinary garden hose can be coiled inside the fermenter to achieve much the same result.

Carbon dioxide gas will be given off during the fermentation. If it is allowed to escape unimpeded, it can carry off a large amount of alcohol vapor. Figure 13-5 shows a fermentation lock available from wine making supply houses. The escaping CO_2 is bubbled through

Figure 13-5: FERMENTATION LOCK

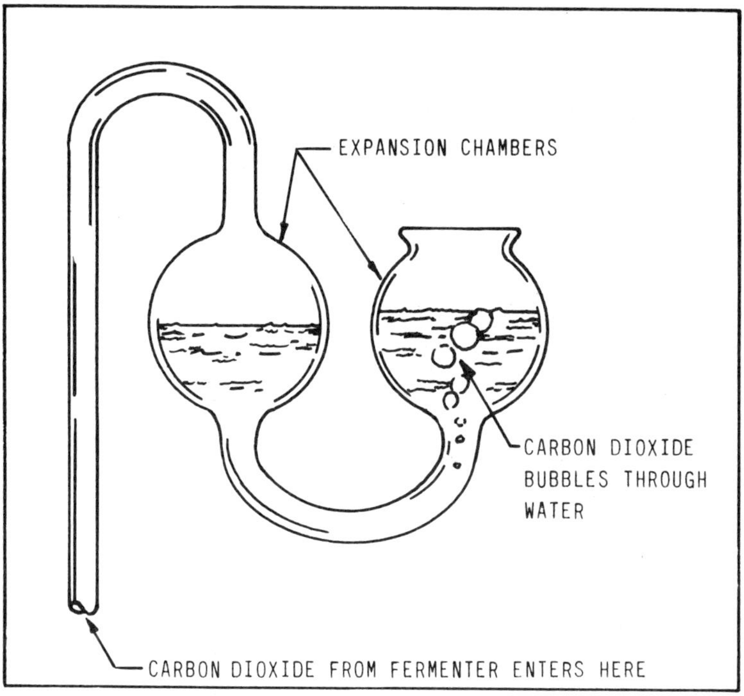

water. The water traps any escaping alcohol vapors. Expansion bulbs keep the water in the trap from either being blown out or sucked back into the fermenter.

Extremely large amounts of gas are given off by large fermenters. The device illustrated in Figure 13-5 should not be used on apparatus larger than about a 5 gallon capacity. Figure 13-6 shows a fermentation lock

Figure 13-6: LARGE FERMENTATION LOCK

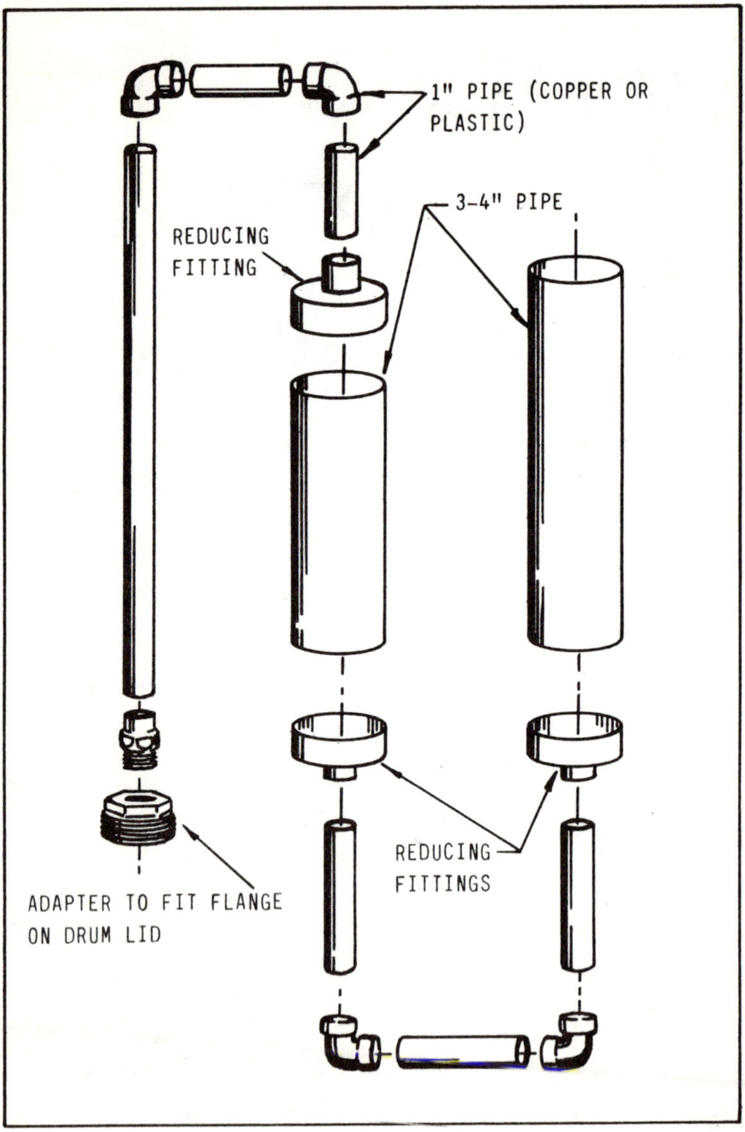

constructed from metal or plastic pipe fittings. Such a unit constructed from 3/4 inch pipe would be adequate for a 50 gallon fermenter. Two inch pipe should be used on a 500 gallon unit, and 3-4 inch pipe on a 1000 gallon fermenter. Remember that pressure can build quickly within the apparatus and, if not allowed to escape, could cause an explosion. Therefore, a safety valve, in addition to the fermentation lock, would be a wise investment.

Continuous fermenters are simply a number of separate fermentation tanks in series. The mash should enter at the bottom of the first fermenter, exit the top, enter the bottom of the second, and so on. Total fermentation time is determined by dividing the total capacity of the fermenters by the flow rate. For example, 1000 gallon fermenter capacity and a 20 gallon per hour flow rate would give a total fermentation time of 50 hours. As noted earlier, sterile feedstock is necessary.

Chapter 14
DISTILLATION EQUIPMENT

DISTILLATION EQUIPMENT

Distillation equipment can be divided into several categories. Simple apparatus, such as illustrated in Figure 11-1 can be constructed on a large scale. However, such a simple "pot still" would achieve less than 100 proof on the first run from an average beer and many separate distillations would be necessary to achieve 190 proof or 95% alcohol. For the production of fuel alcohol such a unit is not economical either in terms of energy input or labor. The only exception would be if the alcohol were used in an injection system where low proof is acceptable.

Rectifying columns, on the other hand, can achieve 190 proof on the first run. Stills incorporating rectifying columns can be designed to run either on a batch or continuous basis. A batch operation still is simply a reflux rectifying column attached to a suitable boiler. The boiler is loaded with beer, the alcohol is distilled out of the beer, and then the still is shut down while the stillage is emptied from the boiler and a fresh batch of beer run in. Continuous operation stills, on the other hand, do not have to be shut down periodically. They can be operated 24 hours a day and, with proper automatic controls, require very little attention.

The basic packed reflux column described in Chapter 11 is only efficient up to a certain point. Figure 14-1 illustrates a cross section of a larger apparatus that operates on the same principle. Here, ascending vapors rise through holes in a perforated plate. The descending liquid flows downward from plate to plate through down-pipes. The liquid does not flow through the holes in the plate because of the pressure exerted by the

Figure 14-1: CROSS SECTION of LARGE COLUMN

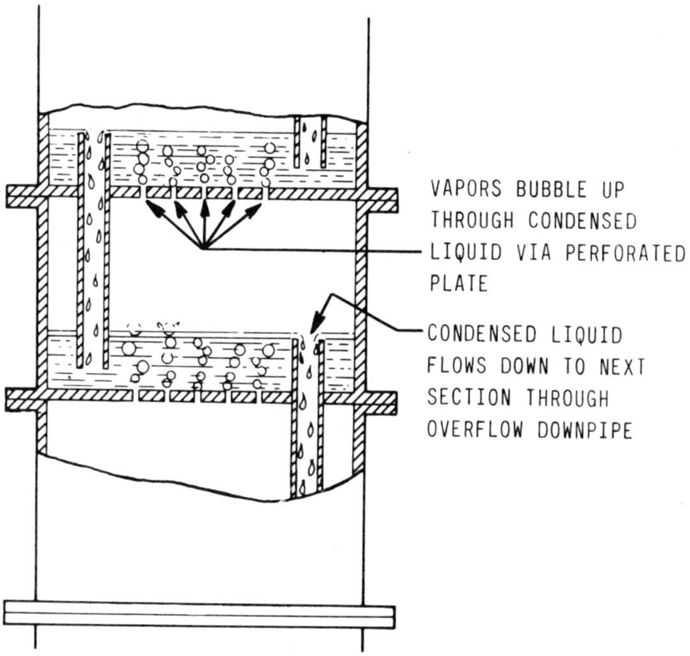

ascending vapor. Thus, a certain amount of liquid is "trapped" on each plate and, as the vapors bubble through it, alcohol is removed from the descending liquid. The effect is similar to the packed reflux column in that a separate "distillation" is performed at each plate. The plate design shown is only one of several possibilities. Various forms of bubble caps, for example, are used on larger columns.

Figure 14-2: CONTINUOUS STILL

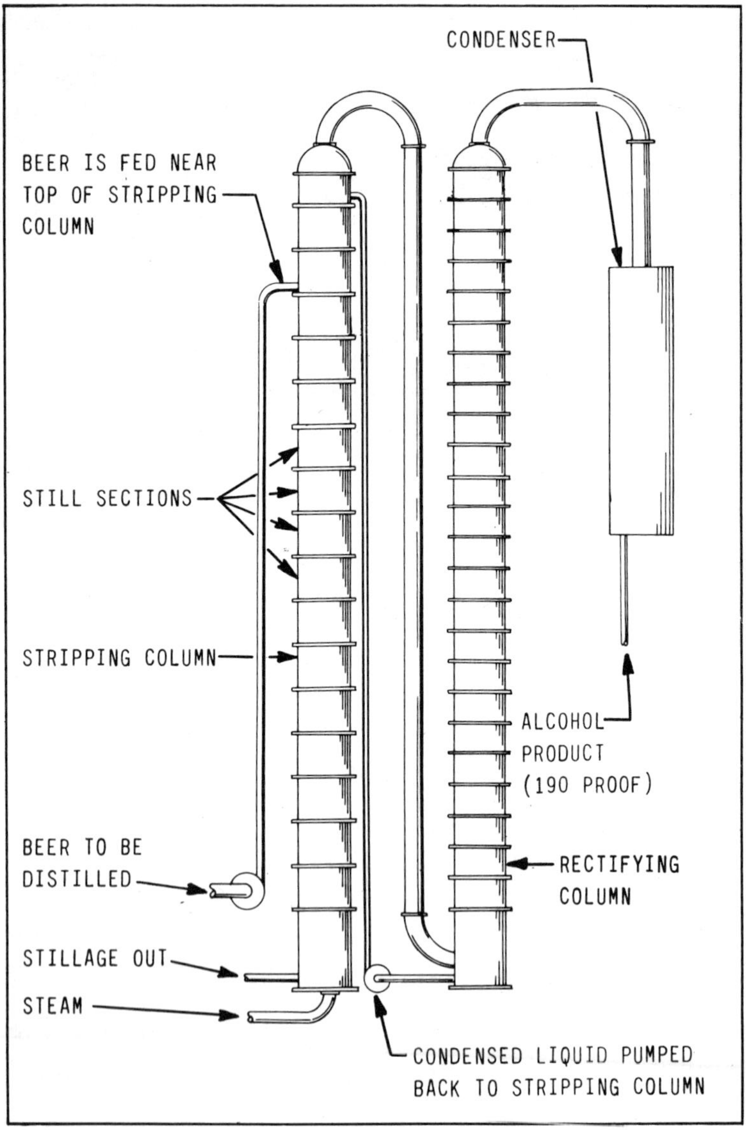

Figure 14-2 is a continuous distillation apparatus incorporating the plates described in Figure 14-1. The apparatus illustrated consists of two columns, although there is no reason why the unit could not be constructed as a single column. The liquid to be distilled (beer) is pumped into the first column near the top. Steam is piped in at the bottom. As the beer descends, the steam "strips" it of alcohol. The alcohol vapors pass over to the next column, and the alcohol-free liquid (stillage) exits from the base. The next column also contains plates similar to those in the first. In this column the alcohol vapors are stripped of most of the remaining water (or "rectified") and exit as 190 proof (95%) alcohol.

The distillation equipment described so far uses the principle of adding heat to boil the beer and provide vapor for the distillation process. Alternately, vapor can also be produced by reducing pressure. In a vacuum, it is easily possible to boil ice water at 32° F. Similarly, alcohol/water mixtures can also be boiled at "room" temperature and below simply by reducing pressure. The equipment consists of a vacuum pump, condenser, and a still pot built to withstand the external pressure created by the vacuum. Although the energy required to run the vacuum pump is probably equal to the amount of energy required to operate a conventional still, this type of equipment merits consideration.

The selection of distillation equipment is largely a matter of economics. Continuous operation stills must be properly engineered and are costly to construct. However, the advantages of automated operation and low labor requirements make them very attractive. For operations producing a large amount of fuel, a continuous still clearly makes sense. Batch stills, although labor intensive, can be built by the layman with relative ease and for a small amount of money. The balance of this chapter describes the construction and operation of such equipment.

Note that solar energy can be used in several ways to provide heat for the distillation process. Solar stills are discussed in

Chapter 15.

SIMPLE REFLUX COLUMN

Figure 14-3 shows a simple rectifying reflux column. The column is simply a length of pipe filled with a packing material to provide a large internal surface area. Aside from the pipe to hold the packing, some sort of screen or retainer is needed at the base of the column to keep the packing from falling into the boiler. The thermometer at the top of the column is necessary to check the temperature of vapors going to the condenser.

A 3 inch diameter column of this design should be about 4 feet long. It would be capable of producing about 1 gallon per hour, depending on the initial concentration of the beer. Similarly, a 4 inch column should be about 6 feet long and should deliver about 2 gallons per hour; a 6 inch column should be at least 10 feet in length and should be capable of 5-6 gallons per hour. Columns of this design do not work well in diameters above 6 inches.

Figure 14-3: SIMPLE REFLUX COLUMN

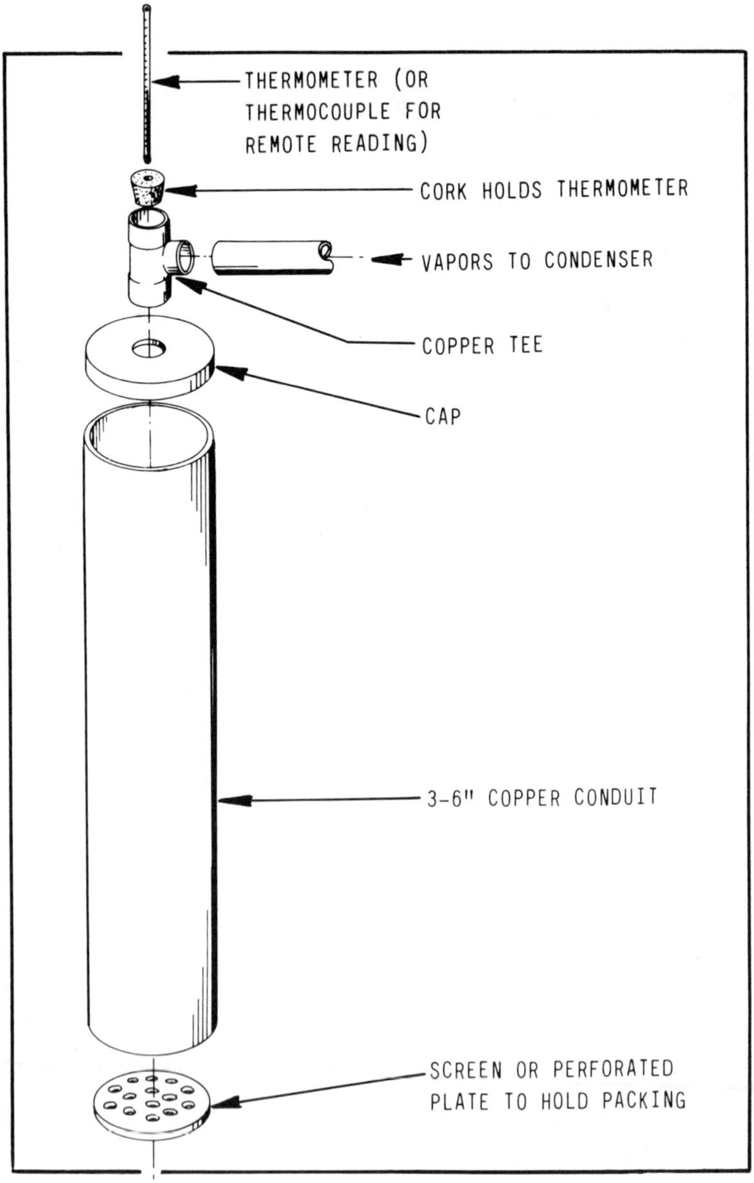

CONDENSERS

The top of the column will have to be connected to a condenser to cool the vapors back into the liquid form. The condenser can be a coil of soft copper pipe inside a suitable container as illustrated in Figure 14-4. Here water is used as the heat exchange medium. On small stills, air cooled condensers are also possible. An old automobile radiator should work very well. The main thing is that the condenser be large enough to cool all of the vapors from the still below 100° F. and, preferably, to about 60° F.

Also, if the vapors going into the condenser are impeded in any great degree, pressure could build up inside the column and boiler. Therefore, the diameter of tubing in a condenser for a 3 inch column should be no smaller than 3/8 inch diameter. For a 4 inch column, 1/2 inch tubing would be the absolute minimum, and for a 6 inch column, the minimum would be 3/4 inch diameter. Note that the effective diameter of a condenser can be increased by connecting two or more

Figure 14-4: CONDENSER CONSTRUCTION

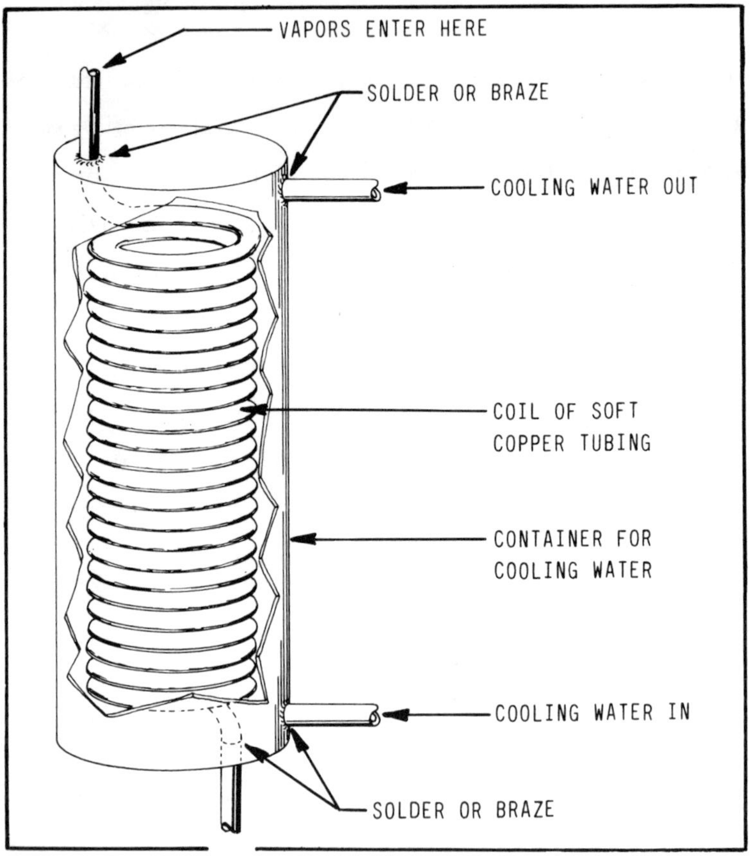

condensing coils in parallel.

BOILERS

To complete the basic apparatus, you will need a boiler or "still pot". A 55-gallon drum can be adapted to this purpose. The drum is best placed on its side to allow maximum surface area for heating and the production of vapor. Alternately, a small column could be attached to the cooking and mashing drum illustrated in Figure 13-1. A more efficient boiler could be made from a hot water heater. Hot water heaters are available that can be fired with wood, coal, electricity, or gas. They are usually glass lined and insulated. Serviceable units can often be found in a junk yard and rebuilt. Other types of boilers can also be adapted for use as a still pot. An example might be certain types of home or commercial hot water furnaces.

REFLUX CONTROL

Successful operation of the simple reflux column described earlier depends on careful regulation of the amount of vapor going to the column. However, it is often difficult to control the amount of vapor produced by boilers that are fueled by wood, coal and similar materials. The reflux control coil illustrated in Figure 14-5 can be added to the column and used to regulate temperatures at

Figure 14-5: REFLUX CONTROL COIL

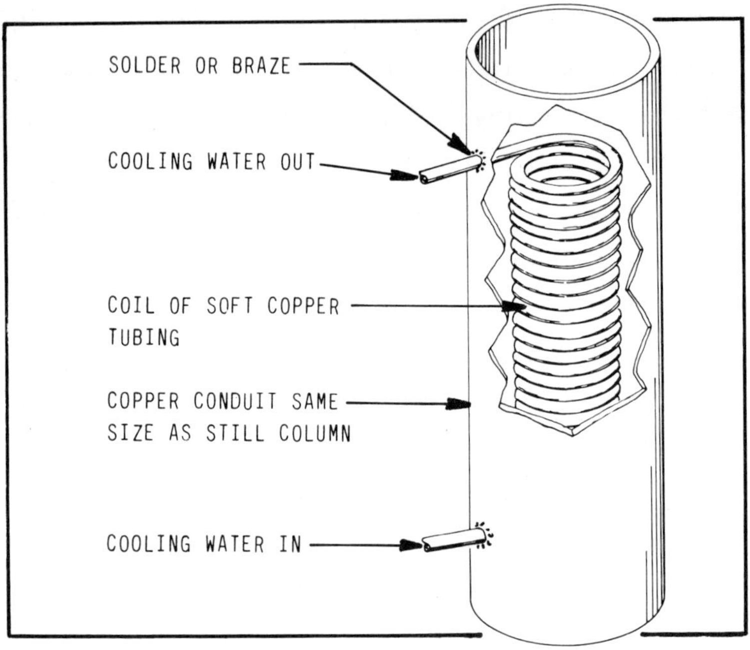

the still head with a great degree of accuracy. The reflux control unit is simply a condensing coil placed in the column and used to control the amount of reflux. It is constructed by wrapping soft copper tubing around a suitable form. The copper coil is then placed inside a section of the column.

In use, cooling water is circulated through the coil to condense a portion of the ascending vapors and, thus, increase the amount of reflux. Adjustment of cooling water in the reflux coils must be very precise. Small needle valves designed for precision metering of liquid should be used. Semi-automatic operation of the still could be achieved by replacing the cooling control needle valves with solenoid valves controlled by temperature sensors within the column.

HYDROMETER SUMP

There are several improvements that can be added to the basic still. One of them is a hydrometer sump as illustrated in Figure 14-6. It is constructed of ordinary pipe and fittings. In use, the product from the condenser is admitted at the bottom and flows out of the top. The hydrometer and thermometer allow a constant check of the proof of the alcohol being produced by the still. The valve at the bottom of the sump allows the unit to be drained.

CONSTRUCTION OF A REFLUX COLUMN

Figure 14-7 illustrates a reflux column

Figure 14-6: HYDROMETER SUMP

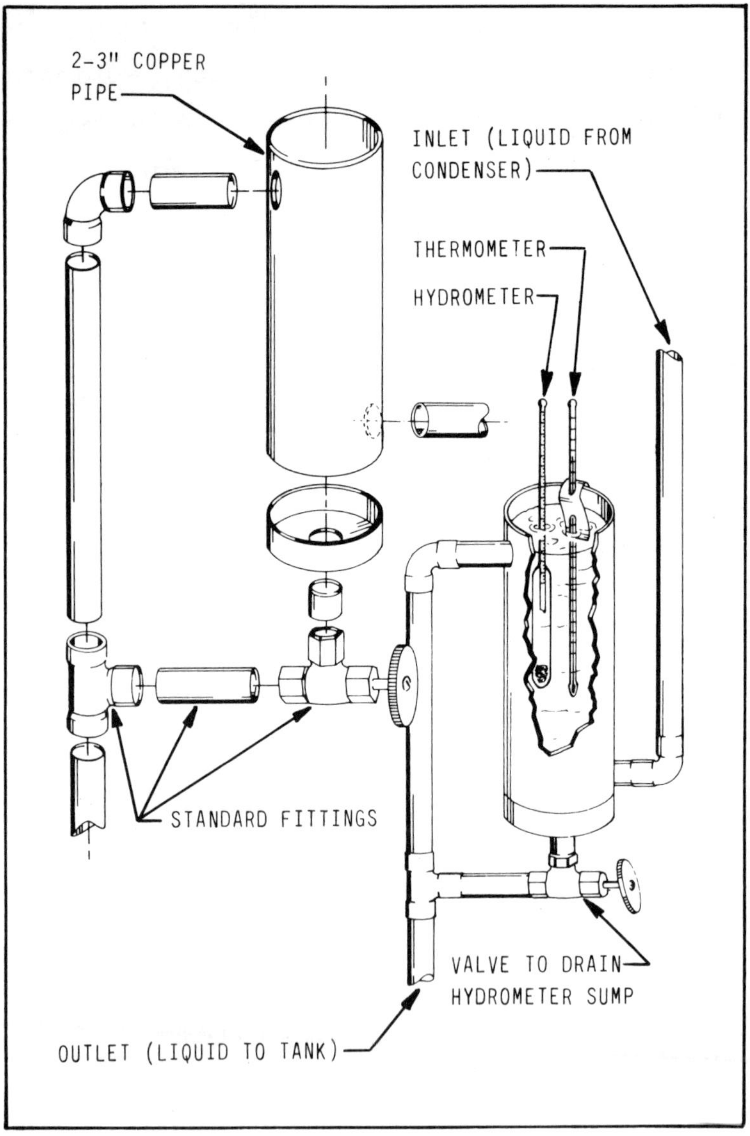

incorporating 2 reflux control coils, a hydrometer sump, and related plumbing. The layout of the condenser, pipes, valves, etc. is for clarity in the illustration and not necessarily the best configuration.

The column can be constructed from copper, iron, or steel pipe and fittings. Aluminum is not suitable because it can react chemically with the alcohol. The use of certain types of rubber and plastic that are attacked by the alcohol should also be avoided. Copper pipe and fittings, although expensive, are easy to work with and are recommended.

To construct, for example, a 4 inch column, begin by cutting a 6 foot length of pipe for the rectifying section, a 2-3 foot length for the stripping section, and 2 lengths about 1 foot long for the reflux control sections. Note that the use of two reflux control sections allows very precise control of temperatures within the column. However, if an easily controlled heat source, such as gas or electricity, is used for the boiler, the lower coil can be eliminated. The reflux control

Figure 14-7: IMPROVED REFLUX COLUMN

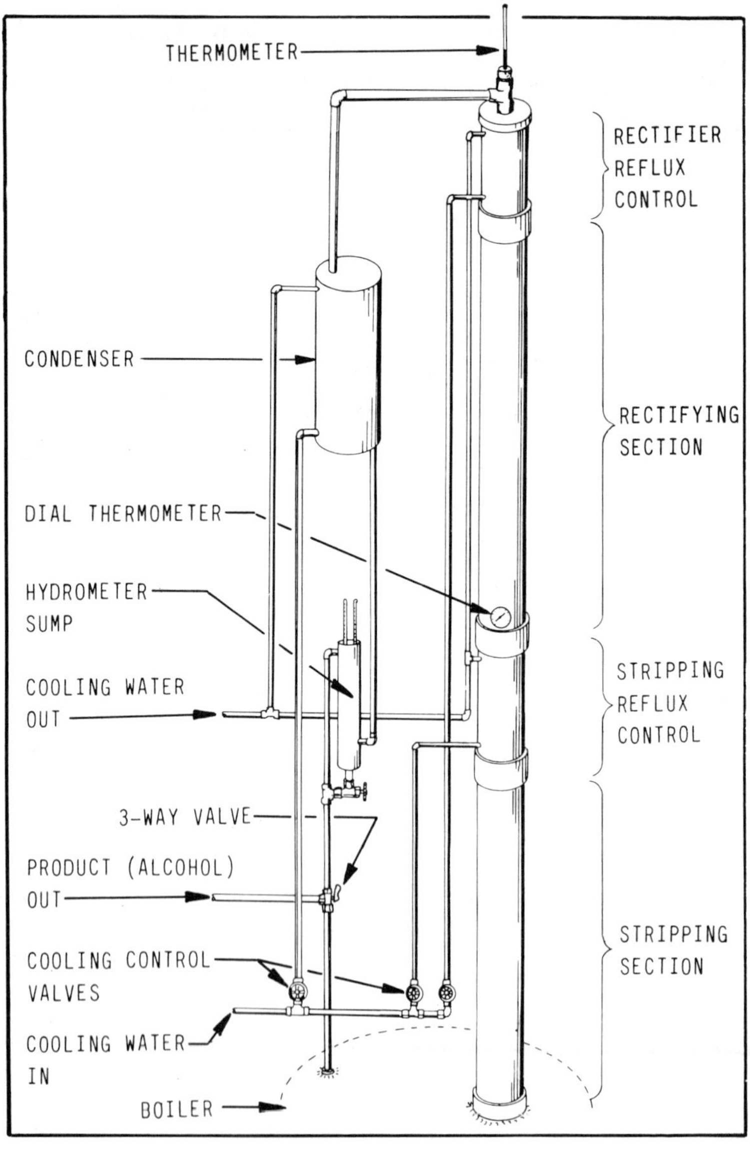

coils are soft copper tubing, 3/16 to 3/8 inches in diameter wound around a suitable form and placed inside the short sections of pipe.

Both the rectifying and stripping sections should be packed. The packing can be marbles, pebbles, broken glass, short pieces of metal or glass tubing, or whatever. Anything that won't rust or react with the alcohol will work. However, the best packing for this type column is probably copper or stainless steel scouring pads. Ordinary steel wool will not work because it will quickly rust. Some sort of screen or retainer is needed at the base of each packed section to keep the packing material from falling into the boiler. The simplest retainer is a section of coarse screening cut to the column diameter and soldered into place.

Once packed, the entire unit can be assembled. The drawing shows the sections joined by pipe couplers.

Figure 14-3 shows the thermometer (or other

temperature sensing device) being held in place by a cork. This is intended as a safety device. Excessive pressure will pop the cork. If some other arrangement is used to hold the temperature sensor, a pressure relief valve, such as used on hot water heaters, should be added either in the column or the boiler.

A worthwhile improvement would be the use of remote sensing thermometers (thermocouples or thermistors) with the read-out located in some convenient place, for example, near the reflux control needle valves. As mentioned eariier, the use of solenoid valves, controlled by remote temperature sensors, in place of (or preferably in conjunction with) the reflux control needle valves would allow semi-automatic operation. A thermometer to measure vapor temperature in the boiler (not shown) is also necessary.

The three-way valve located below the hydrometer sump allows the impure product to be run back into the boiler for redistillation. In addition, two collection tanks, one for high proof product and one for low proof residue, are needed. As will be explained in the section on still operation, the low proof tank should be set up so that its contents can be run back into the still pot at the beginning of each successive run.

OPERATION OF THE STILL

In operation, the boiler or still pot is filled to no more than the 3/4 level with the beer to be distilled. As the liquid begins to boil, vapors rise in the column. After a while, the column will come up to temperature, and an equilibrium will be established. For a normal beer concentration of about 8%, the initial temperature of the vapors in the still pot will be about 200° F. The vapors will cool as they rise in the column and, at the top, they should stabilize at 173° F. which is the approximate boiling point of the water/alcohol azetrope.

If the boiler is supplying more vapor than the column can handle, the temperature at the still head will rise above 173° and the proof of the product going to the condenser will be reduced. Conversely, if the boiler is not producing enough vapor, the temperature at the still head will be low and no vapors will be going to the condenser. On a still without reflux control coils, the heat source at the boiler must be adjusted to keep the amount of vapor within the range that can be handled by the column. On stills with reflux control, the boiler is adjusted to produce an excess amount of vapor. The lower reflux control coil is adjusted so that the thermometer located just above it reads 180° F. and the top reflux control coil is adjusted to maintain the still head temperature at exactly 173° F.

After the column has stabilized, the 3-way valve is opened to let the product flow to the high proof tank. Note that when you begin a distillation run, certain low boiling vapors may come over, and a small amount of liquid may come out of the condenser that is not ethanol. This liquid is composed of substances in the beer that have a lower boiling point than ethanol. However, they do not affect fuel value and, for all practical purposes, can be ignored.

As the distillation progresses, the vapors in the still pot will contain a greater and greater percentage of water and a correspondingly lesser proportion of alcohol. The still pot vapor temperature will rise. Eventually, a point will be reached where there is too little alcohol in the vapor for the column to achieve effective separation. The temperature at the still head will rise slightly and the proof of the product will be lower. At this point, it is best to collect the product coming from the condenser in the low proof container mentioned earlier.

The distillation should be continued until the temperature at the still head equals the temperature of the vapors in the boiler, which will be near 208-212° F. depending on altitude, atmospheric pressure, and the amount of dissolved material in the beer. Note that when switching to the low proof phase of the distillation, the reflux control coils should be turned off and the column temperatures allowed to rise naturally.

When all the alcohol has been removed from

the beer, as indicated by boiler and still head vapor temperatures, as mentioned before, the distillation is complete. The beer, now called "stillage", is drained from the boiler, and a fresh charge of beer run in. The low proof "tail" from the previous run is added to the fresh charge of beer, and the still is ready for another run.

CAUTION

There are several inherent dangers in the construction and use of the equipment described in this chapter. Alcohol, and alcohol vapors, are flammable. Mixtures of alcohol vapors and air can be explosive. The equipment should be located in an area that receives adequate ventilation, and preferably outdoors or in a simple shed away from other buildings. The still should be electrically grounded to prevent the buildup of static electricity. Above all, reasonable care and common sense should be exercised. If you are not sure about something that could be potentially dangerous, find out before you proceed!

Chapter 15 SOLAR STILLS

GENERAL DISCUSSION

Solar stills can be simple or elaborate. The solar analogy to a simple pot still is easy to construct and operate. Solar stills that incorporate stripping and reflux sections are also possible, but they are difficult to operate without sophisticated instrumentation and controls. This is because there is a delicate balance between temperature, feed rates, and the amount of vapor in the column. The slightest variation in the input of solar radiant energy will destroy this balance. To operate complex solar stills, all of the variables must be sensed and controlled within very close tolerances. This, unfortunately, means expensive electronic sensors, controls and valves. The overall efficiency of a complex solar still is low and, to offset the cost of instrumentation, the still must be very large.

It is usually better, for larger installations, to use solar power indirectly. For example, a solar installation could be used to preheat the water used for cooking or a solar boiler could be used to supplement a regular boiler for the production of steam.

Simple solar stills have the disadvantage of producing low proof alcohol. However, they are entirely adequate when used to produce fuel for an injection system. An example of a simple fermentation set-up and solar still to produce fuel for an alcohol injected engine is described in Chapter 17. In this application, the solar still makes a lot of sense.

PRINCIPLE OF OPERATION

Figure 15-1 illustrates a solar still of the type used in survival gear to distill fresh

Figure 15-1: BASIC SOLAR STILL

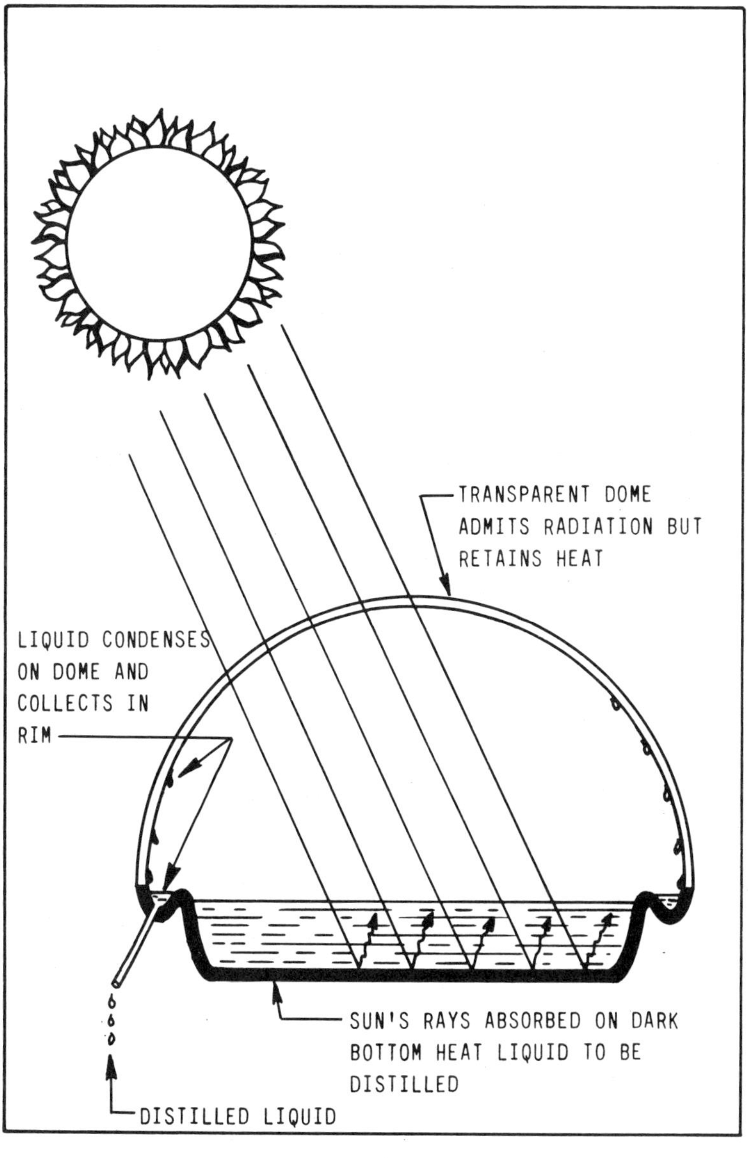

water from sea water. The sun's rays pass freely through the transparent dome and are absorbed by the dark bottom of the still. The liquid in the bottom of the still is heated. Vapors rise, contact the inside of the transparent dome (which remains relatively cool), and condense. The distilled liquid collects in the trough around the rim and is collected through the attached tube.

The still illustrated could just as easily distill alcohol as water. Such stills are available from various suppliers and are fairly inexpensive. However, before purchasing one of these units make sure that the alcohol vapors will not damage whatever plastic is used in the still's construction.

CONSTRUCTION OF SOLAR STILLS

Figure 15-2 shows a cross section of an easily constructed solar still. The box is made of plywood and is about 6 inches deep. The overall height is 2-3 feet. It can be constructed in any convenient length. It is important that the box be vapor tight.

Figure 15-2: PASSIVE SOLAR STILL

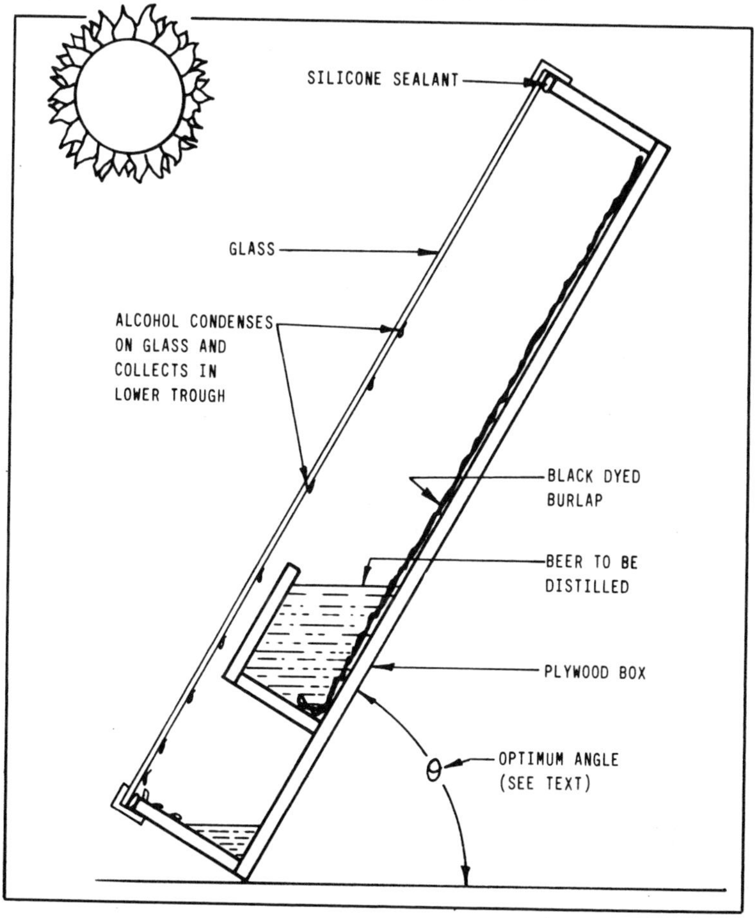

Otherwise, the alcohol vapors will escape. Therefore, the inside of the box should be sealed and painted with flat black, chemical resistant, epoxy paint. The glazing should be sealed with a gasket of silicone. A cheaper alternative to glass is translucent fiberglass used in the construction of greenhouses. It is available from most building supply houses in rolls that are 48 inches wide. A valve should be provided to fill and drain the beer trough. The drain opening in the lower alcohol trough must be left open to relieve internal pressure. To prevent uncondensed vapor loss, a simple "U" trap, as illustrated in Figure 15-3, can be installed. The optimum angle for the box as far as the sun is concerned is your latitude plus 15° in the winter and minus 15° in the summer. In the spring and fall the optimum angle is equal to your latitude.

In operation, the beer travels up the black-dyed burlap by capillary action, the alcohol evaporates, condenses on the glass, and collects in the lower trough. The temperature in the box is self-regulating or "passive" because as the box heats up, more alcohol will evaporate. This has a cooling effect. Conversely, if the temperature in the box drops, less alcohol will evaporate, and the temperature will rise. The vapor temperature inside the box should, therefore, remain relatively stable. Typically, the

Figure 15-3: VAPOR TRAP DRAIN

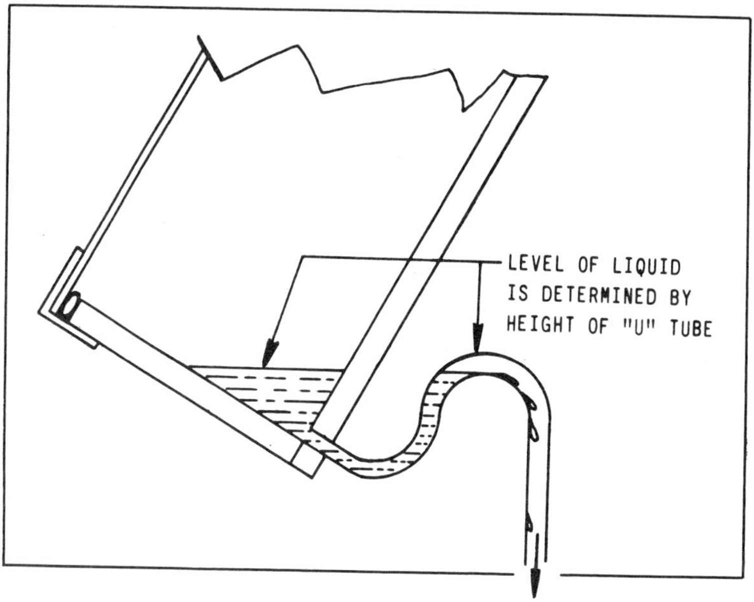

initial temperature will be about 175° F. and the proof of the distillate will be about 100-120, depending on the alcohol concentration of the beer. As the alcohol is removed from the beer, the temperature inside the tank will rise and the proof of the alcohol collected will decrease. Therefore, after about 1/2 to 2/3 of the alcohol has been collected, the distillate can be run into a second container for redistillation with the next run.

With slight modification the still in Figure

15-2 can be converted from passive batch operation to a more or less continuous operation. A drip tube is installed across the top of the still. Instead of the beer flowing up the burlap by capillary action, it is allowed to drip down the burlap from above. Figure 15-4 shows a solar still set up for this type of operation. Notice that the height of the still is much greater, about 8 feet. The object is to have all of the alcohol evaporated from the beer by the time it reaches the lower trough. This is not always possible. Therefore, some means of recirculating the beer usually must be provided. Control of the temperature within the still is no longer completely "passive". To a greater extent, it must be "actively" controlled by adjusting the flow rate of the beer. An increased flow rate will lower the temperature inside the box and a reduced rate will raise the temperature. An alternate to controlling the flow rate would be to install louvers or a shade, activated by a thermal sensor, to control internal temperature. The closer the temperature to 173° F., the higher the proof of the alcohol. Because the sun's energy is not consistent, some means

Figure 15-4: ACTIVE SOLAR STILL

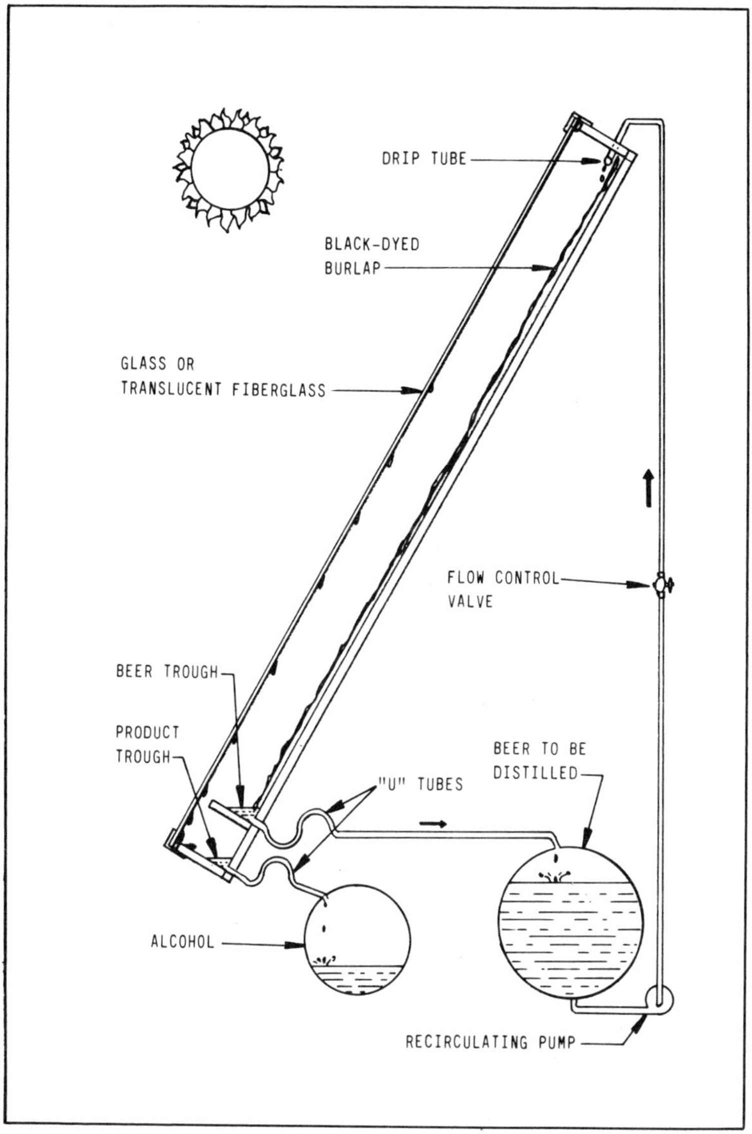

of automatically controlling the flow rate to maintain optimum temperature is almost mandatory.

The passive system has the advantage of being self-regulating as far as temperature is concerned and the disadvantage of producing lower and lower proof as the distillation continues. Because of the fixed volume of the beer trough, it can only process a limited amount of beer at one time. The active system, if properly regulated, is capable of a higher overall proof on the first run and, with a recirculating tank, can process a larger amount of beer. It has the disadvantage of requiring active control of the internal still temperature.

Both stills produce a high enough proof for injection systems. However, if the fuel is to be burned in place of gasoline, the alcohol must be redistilled to achieve an acceptable proof. There is no reason, however, why several stills can not be connected in series with the product from the first being fed to the second, and so on. Also, although overall efficiency is low, the energy from the sun is "free" and stills of this design can be built as large as desired.

Chapter 16
GOVERNMENT REGULATIONS

FEDERAL REGULATIONS

Federal law (Internal Revenue Code of 1954, 26 U.S.C. 5171) states: "Every person shall, before commencing the business of a distiller, make application for and receive notice of the registration of his plant". This means, simply, that you must register your plant and get your permit before you begin producing alcohol! Violation is a federal offense punishable, possibly, by a jail sentence.

The Bureau of Alcohol, Tobacco and Firearms of the U.S. Treasury Department has the responsibility of administering the laws concerning ethanol production. They maintain several regional offices throughout the country and the addresses are listed in the appendix. The ATF Bureau has considerably relaxed and streamlined the process for registering small alcohol plants. For example, until recently the minimum bond requirement was $10,000.00. Now it is only $100.00 for plants producing less than 2500 gallons per year.

There are two basic types of alcohol permits. The first is an "operating" permit that allows sale and distribution of the (properly denatured) product. The second is an "experimental" permit which does not authorize sale of the alcohol. Qualifying for an operating permit is an extensive procedure that requires numerous forms and an investigation to determine whether operation of the plant can be legally approved. However, qualification of an experimental plant is a simpler process.

To apply for an experimental permit, write to the nearest regional office of the ATF Bureau. Your letter must completely describe: (1) all land, buildings and equipment (stills, fermenters, tanks, etc.) to be used in producing alcohol; (2) the planned uses of the alcohol (use in tractors, lawnmowers, automobiles, etc.); (3) the property on which vehicles using the alcohol as fuel will be operated; (4) the process you will use to produce the alcohol (raw materials, yeast, enzymes, distillation method, etc.); (5) how you will dispose of the waste materials, stillage, etc. (i.e. feed it to livestock); and (6) the type of security you will provide to prevent theft of the alcohol.

Your letter should also state how much alcohol you will produce in an "average" 15 day period. Naturally, if you wish to qualify for the minimum bond, the total yearly amount should be less than 2500 gallons. When your application is approved, you will need to file a distiller's bond. The bond form (ATF 2601) will be sent by the regional office.

Undenatured alcohol may only be used on the "premises" described in your letter. To use the alcohol on public roads or away from your property, it must be denatured as described later in this chapter.

STATE AND LOCAL LAWS

Compliance with Federal law in no way exempts you from compliance with state and local laws. You would be wise to check these before starting to produce alcohol. Pennsylvania, for example, used to require a $2,500.00 bond in addition to the Federal requirements. Happily, this law has been changed. Many other states are also in the process of revising and updating their requirements. However, archaic laws are still on the books in many areas. Letters to your state and local representatives could help to get them changed.

Also, many of the laws concerning motor fuel on both the state and federal level were written with gasoline in mind. For example, it may be illegal for you to operate your vehicle without a catalytic converter even though such a device is unnecessary with alcohol. Again, it would be wise to check beforehand and beneficial to write letters to get the outmoded laws changed.

DENATURING

The government is extremely concerned that alcohol intended for non-potable use may be turned into "drinking" alcohol. The present tax on potable alcohol is about $10.50 per gallon, while alcohol, for fuel and industrial purposes, is not similarly taxed. To insure that non-potable alcohol stays that way, it must be "denatured". Denaturing is the addition of certain substances that are difficult to remove from the alcohol, and make it unfit for drinking.

Because ethanol is used for many different purposes, and because the chemicals used to denature the alcohol may interfere with later use of the alcohol, there are different denaturing formulas. The government divides these formulas into two categories. The first, "Completely Denatured Alcohol" (CDA) formulas make it virtually impossible to separate the alcohol from the chemicals used to render it unfit to drink. CDA alcohol can be handled and sold without restriction. The second class of formulas, "Specially Denatured Alcohol"

(SDA), have restrictions as to the handling and storage of the alcohol.

Most of the CDA formulas call for the addition of chemicals such as ethyl acetate, methyl alcohol, or methyl isobutyl ketone. These chemicals must be purchased. CDA formula #19, for example, calls for the addition of 4 gallons of methyl isobutyl ketone and 1 gallon of either kerosene or gasoline to every 100 gallons of alcohol. On the other hand, SDA formula #28-A calls only for the addition of one gallon of gasoline per 100 gallons of alcohol. This is accomplished naturally if you are making gasohol. However, since it is an SDA formula, you will have to have approval from the ATF Bureau and comply with the restrictions for its storage and use.

Before purchasing special denaturing chemicals, write to the regional ATF Bureau office. They might approve an SDA formula such as #28-A if you agree, for example, to install a locking gas cap on your vehicle.

Remember, it is illegal to use your alcohol off

your premises unless it is denatured by an approved formula.

Chapter 17
PUTTING IT ALL TOGETHER

LARGE AND SMALL SYSTEMS

VERY SMALL

The author has alcohol/water injection systems installed on his 1978 Honda and 1975 Ford pick-up truck. The injection systems, purchased from Spearco Performance Products, Inc. (address in appendix), are designed for water injection. Consequently, they were modified to (approximately) double the flow rate to the carburetor under peak load conditions. This enables maximum efficiency from a 120 proof alcohol/water injection mixture. Under average driving conditions, the gasoline saving amounts to 25–30%. The alcohol injection mixture is prepared from miscellaneous vegetable matter, damaged fruit, and garbage that is collected (free) from a number of sources. It is fermented directly in two 44-gallon plastic trash cans and distilled in a solar still as described in Chapter 15. The entire set-up, including the injection systems, cost less than $150.00 (in 1978) and the labor requirement is only an hour or two per week. Family gasoline consumption has been cut from an average of 50 gallons per week to about 35. The whole investment was paid back in less than 4 months, and all current savings are pure profit!

SMALL

A nearby farmer, who is semi-retired, processes a variety of agricultural surplus in a 500 gallon cooker that also serves as a fermenter and still pot. One run a week produces about 400 gallons of beer that yields, on an average, 35 gallons of 190 proof alcohol. The alcohol is used in a tractor, the family car, and a one-ton truck. The entire "plant" was built (in 1978) for about $1000.00. The batch still and column (4" diameter) only produce 2 to 3 gallons per hour (at 190 proof), and the total labor requirement is about 20 hours.

MEDIUM

Another farmer in the area has a somewhat larger operation. He processes artichoke culls in a 4000 gallon cooker that also serves as a fermenter. However, distillation is done in a 10 inch continuous column that produces 20 gallons of 180 proof alcohol per hour. Working intermittently when feedstock material is available, 2 batches are cooked and fermented each week. The resulting 6000 gallons of beer are distilled in about 24 hours. Since the still is fully automated, total labor to produce 500 gallons of fuel is about 20 hours. The still was built in the farm machine shop from complete blueprints purchased from the Alcohol Technology Corporation (see appendix) at a cost of about $10,000.00. Of this amount, about $4,000.00 was spent on automatic controls and instrumentation. A comparable distillation unit, purchased commercially, would cost between $25-50,000.00! In addition, the cooker/fermenter was built for about $4,000.00 and miscellaneous plumbing cost another $1,000.00, all at 1979 prices. The planned addition of two 5,000 gallon fermenters will cost another $7-8,000.00. The desired capacity is to produce 15-20,000 gallons of fuel alcohol each fall when the surplus crops are available.

LARGE

An even larger fuel alcohol production facility is currently being constructed by a Mid-West dairy farmer. This facility will process corn. Each 70 pounds of unshelled corn will yield a bushel (56 pounds) of shelled corn and 14 pounds of corn cobbs. The shelled corn, when processed, will yield about 2.5 to 2.7 gallons of alcohol. The corn cobbs, when dried and burned in a boiler, will produce 91,000 Btu which is enough to completely fuel the cooking and distillation operations. The facility centers around a steam batch cooker, several fermenters, and a 12 inch continuous, fully automated, still. Initial production will be in excess of 50,000 gallons per year, although the facility is capable of several times this amount. The stillage is to be fed to a resident dairy herd as a high-protein supplement. The possibility of using manure from the cattle to generate methane is also being considered. The entire system is being professionally engineered and purchased commercially at a cost of about $250,000.00.

The foregoing examples illustrate a number of points. First, that alcohol production is possible on all levels, from the author's modest system to a fairly large "on farm" plant. Secondly, all of the systems described fill a need, all of them are cost efficient, and all will pay back the entire plant cost in a relatively short time. Finally, all of the systems provide a certain degree of independence from petroleum fuels, and all are operated totally from renewable resources.

CONSIDERATIONS

The main areas of cost in setting up and

operating an alcohol plant are: (1) the cost of the physical plant; (2) the cost of raw materials or feedstock; (3) the cost of fuel to run the plant; and (4) the amount of labor involved.

You can begin producing alcohol with a modest capital investment or you can spend a great deal of money. A farm or cooperative might find an alcohol facility a bargain at almost any price. Not only can the installation eliminate the use of petroleum, it can provide independence or "crop insurance" in these perilous times. As for very small systems, they pay for themselves and, if everyone produced even a small percentage of his own motor fuel, there would be no immediate fuel crisis.

Farmers have the advantage of available surplus and damaged crops that might otherwise go to waste. They also have the advantage of combined operations whereby stillage is fed to animals (saving feed costs) and manure is used to generate methane to run the alcohol facility. Others are not quite so fortunate. Nevertheless, alcohol feedstocks are available for very little cost if time is spent to locate them. City dwellers, for example, can purchase damaged materials from wholesale food distributors.

Fuel to run the still and cooking operation is another important cost consideration. The fuel must be cheap and available. Wood, straw, corncobbs and the like are hard to handle but are good sources of heat. Coal might be considered in certain areas as might electricity. Whatever the fuel, though, it should be from a renewable source.

Labor is a consideration closely related to plant cost and the degree of automation. Small batch equipment requires a disproportionate amount of labor for the alcohol produced. The "small" operation previously described requires 20 hours labor for less than 40 gallons of fuel. The "medium" operation produces over ten times the amount of alcohol for the same amount of work. The difference is in the size of the equipment. In the long run, larger equipment often pays for itself in terms of labor and economic operation.

Chapter 18 THE FUTURE

PRESENT TECHNOLOGY

Almost all of the technology currently used in the production of fuel alcohol was developed 30-40 years ago. During the Great Depression, much research was done concerning the use of crop surplus for the production of fuel alcohol. The research was never put into practice partly because of pressure from the oil lobby. Interest in alcohol fuel again arose during the Second World War when oil was in short supply. It is no secret that much of Europe, and the entire Third Reich, ran on alcohol fuel from about 1943 on. After the war, interest in alcohol again declined, partly because gasoline was cheap and available, and partly because of the oil lobby. Today, with a new crisis upon us, this existing fuel alcohol technology is being rediscovered.

It is amusing to note that a lot of the "new" advances being claimed are not new at all. A manufacturer of distillation equipment, for example, claims the discovery of a process of using gasoline as an alcohol drying agent in the direct production of gasohol. In fact, U. S. Patents 1,490,520 (1924), 1,699,355 (1929), 1,744,504 (1930), 2,371,010 (1945), and 2,591,672 (1952) cover the subject quite well! Soon, automobile manufacturers (after much research and the resulting tax writeoffs), will "discover" that alcohol can be used in internal combustion engines, and tractor manufacturers will "remember" that they have been exporting alcohol fueled farm equipment to many foreign countries, principally in South America, for the past 50 years.

The point is that most of the solution of the energy problem regarding alcohol fuel is already in existence, waiting to be re-discovered and put to use.

NEW TECHNOLOGY

The current (and hopefully continuing) interest in alcohol fuel is generating new research and many promising developments are at hand. The following is a small portion of what is being done.

IMMOBILIZED ENZYMES

One of the most important areas of current research concerns the enzymes used to convert starch to sugar. Enzymes are expensive and are a major expense in starch conversion. Using present technology, enzymes can only be used once. They go into solution in the mash and are all but impossible to recover for reuse. The current research (begun in the 1950's) involves immobilizing the enzymes. Instead of being allowed to go into solution, the enzymes are fastened or "immobilized" on an inert substrate. The mash is then passed over a bed of these immobilized enzymes. The end result is that the enzymes do not go into solution and they may be reused many times at a considerable saving. This technology is not yet fully developed for starch conversion, but should be available in the near future.

CELLULOSE CONVERSION

Cellulose is the most plentiful and cheapest potential source of fuel alcohol. Attempts are being made to develop new enzymes to economically convert cellulose to glucose. The problem, as mentioned earlier, is freeing the cellulose from the lignin. Strains of organisms, such as those that cause dry rot in wood, are being studied. The object is not only to extract the cellulose, but to allow easy recovery of the lignin. Lignin, in turn, makes an excellent fuel with which to run a still.

ALTERNATIVES TO DISTILLATION

The entire distillation process is also under scrutiny. The research is in three basic areas. First is the development of large, efficient solar distillation apparatus. The approach is either to develop a still that will completely strip the beer and deliver high proof alcohol, or to use solar energy to generate steam to run a conventional still. Another promising area involves elimination of the distillation process altogether. One method is the use of membranes that will selectively pass the alcohol but retain the water and other impurities in the beer. The third area of research involves extraction of the alcohol with solvents. So far, these processes involve the use of heat and a certain amount of redistillation. Hopefully, the perfected system will consume less energy than conventional distillation.

BIOLOGICAL RESEARCH

It is known that plants must first produce sugar in order to make starch or cellulose. If there were some way to pick plants at the peak of sugar production, the manufacture of ethanol could be greatly simplified. In addition, crops with inherently high sugar content could be mutated to produce even more sugar and grown specifically for the production of alcohol.

There are certain bacteria, algae, and fungi that can ferment alcohol directly from cellulose without a conversion step. It is also within the realm of possibility that a plant could be developed to produce alcohol directly. Certain plants, such as mint, already possess the basic biological mechanism! Even the relatively simple development of a yeast strain that is more tolerant to alcohol would allow more concentrated mashes and a significant saving in distillation energy.

CONCLUSION

The full utilization of existing technology and the promise of new developments will make the production of alcohol fuel easier and more economical in the near future. However, as fossil fuel supplies dwindle, it will become increasingly important to utilize every shred of available material and waste in the production of energy. Aside from the large scale production of alcohol fuel, self-contained, automatic appliances that could turn all sorts of waste material into useable fuel would be an important development. Electric vehicles and small, regional hydro-electric plants would also help, as would full utilization of solar, geothermal, and other energy alternatives.

It is important to realize that the energy problem will not solve itself. It is already the "eleventh hour" and nothing short of a concerted, world-wide effort will suffice!

APPENDIX

ATF BUREAU REGIONAL OFFICES

U. S. Department of the Treasury, BUREAU OF ALCOHOL, TOBACCO, and FIREARMS may be contacted at the following regional offices:

CENTRAL REGION (Indiana, Kentucky, Michigan, Ohio, West Virginia): 550 Main Street, Cincinnati, Ohio 45202

MID-ATLANTIC REGION (Delaware, Washington D. C., Maryland, New Jersey, Pennsylvania, Virginia): 2 Penn Center Plaza, Room 360, Philadelphia, Pennsylvania 19102

MIDWEST REGION (Illinois, Iowa, Kansas,

Minnesota, Missouri, Nebraska, North & South Dakota, Wisconsin): 230 South Dearborn Street, 15th Floor, Chicago, Illinois 60604

NORTH-ATLANTIC REGION (Connecticut, Maine, New Hampshire, Massachusetts, New York, Rhode Island, Vermont, Puerto Rico, Virgin Islands): 6 World Trade Center, 6th Floor, P. O. Box 15, Church Street Station, New York, New York 10008

SOUTHEAST REGION (Alabama, Florida, Georgia, North & South Carolina, Mississippi, Tennessee): 3835 Northeast Expressway, P. O. Box 2994, Atlanta, Georgia 30301

SOUTHWEST REGION (Arkansas, Colorado, Louisiana, New Mexico, Oklahoma, Texas, Wyoming): Main Tower, Room 345, 1200 Main Street, Dallas, Texas 75202

WESTERN REGION (Alaska, Arizona, California, Hawaii, Idaho, Montana, Nevada, Oregon, Utah, Washington): 525 Market Street, 34th Floor, San Francisco, California 94105

INJECTION SYSTEMS

SPEARCO PERFORMANCE PRODUCTS, INC., 10936 La Cienega, Inglewood, California 90304 markets an excellent water injection system that is easily converted to alcohol/water mixtures.

M & W GEAR COMPANY, Route 47 South, Gibson City, Illinois 60936 markets an injection system for turbo-diesel tractors.

OTHER SOURCES of injection systems can be located through your local auto parts outlet or speed shop.

DISTILLER'S YEAST

UNIVERSAL FOODS CORPORATION, 433 East Michigan Street, Milwaukee, Wisconsin 53201 markets Distiller's Active Dry Yeast (DADY).

ENZYME SUPPLIERS

G. B. FERMENTATION INDUSTRIES, INC., 1 North Broadway, Des Plaines, Illinois 60016 markets enzymes for starch conversion.

MILES LABORATORIES, Enzyme Products Division, P. O. Box 932, Elkhart, Indiana 46515 markets enzymes for starch conversion.

NOVO LABORATORIES, Enzymes Division, 59 Danbury Road, Wilton, Connecticut 06897 markets enzymes for starch and cellulose conversion.

PLANS AND BLUEPRINTS

MOTHER EARTH NEWS, P. O. Box A, East Flat Rock, North Carolina 28726 sells plans for 3 and 6 inch batch stills and related cooking and fermentation equipment.

SOLSTICE DESIGNS, INC., P. O. Box 2043, Evergreen, Colorado 80439 sells plans for 6, 12 and 18 inch continuous stills and related equipment. They also offer complete technical and consulting services.

GENERAL INFORMATION

GASOHOL U.S.A., Box 9547, Kansas City, Missouri 64113 is an excellent magazine devoted to fuel alcohol. They began publishing in June 1979 and a collection of back issues would constitute a good reference library. Subscriptions are $12.00 per year and back issues are $1.50 each.

THE ALCOHOL FUEL INFORMATION CENTER, c/o Dr. Richard L. Spencer, Southwest State University, Marshall, Minnesota 56258 provides a packet of information on alcohol production, including suppliers, still manufacturers, and so on.

OTHER SOURCES include the U. S. Department of Agriculture (volumes of information were published during 1930-50), your state agriculture service, and your local library.